Marco the Great and the History of Numberville

By

SK Bennett

Illustrations by Rylee Heavner

Marco the Great and the History of Numberville

Copyright © 2023 by SK Bennett

All rights reserved.

Published by:

Blue Café Books *for*
www.carladupont.com
Atlanta, GA

Printed in the USA.

ISBN: 979-8-9880861-0-9

Credits
Editorial: Carla DuPont
Cover Design: Garrett Myers

TO A WHO ALWAYS THOUGHT NUMBERS WERE THE ENEMY AND TO R WHO WAS A MASTER OF NUMBERS BUT NEVER THOUGHT VERY MUCH OF THEM.

CONTENTS

MARCO THE GREAT

AND THE HISTORY OF NUMBERVILLE

PREFACE

Many believe that in the beginning there was nothing. They are right. However, nothing wasn't alone. Nothing (their actual name is Zil) was surrounded by dust – tiny particles that spanned forever and ever.

Zil enjoyed the solitude of the massive empty space. There was always somewhere new to explore, always more dust to collect, and for a long time (time always existed) Zil did just that. Eventually, an urge began to grow within the nothing. Zil longed to know why they existed, what they were, and with only a simple thought, Zil created the very thing that would govern all of existence – change. Not content to keep things as they were, Zil used all their force to begin pushing together the particles of dust they had collected – forming it into something new. The result was what would be known as the Mirror of Wonders.

When it was finished, Zil peered into the face of the Mirror and saw themselves – saw nothing. Outraged, Zil screamed. "If I am not anything, why do I exist?! What is my purpose?"

Surprisingly, the Mirror responded.

Nothing is powerful, for how could you know if there is something without nothing?

It wasn't until much later that Zil realized what they had created. The Mirror, being formed from the dust of all of time and space, represented the mightiest creature to ever exist – true knowledge.

The Mirror instructed Zil to first form a scale – for without a way to measure and compare, Zil would never understand their own creations. Zil obeyed, spending eons gathering the dust needed until ultimately the Great Scale was born. As Zil looked over their invention, they knew it was good and that the scales would forever ensure balance throughout the universe.

Soon, you will hear all about the amazing and mind-boggling events that happened next. But, we must warn you. Just as Ptolemy assumed Earth was at the center of the Universe, humans have an uncanny ability to make everything about themselves. While Zil used the scales to create fantastical magical beings who would rule over all of time and space, exactly how many is under debate. Our history assumes that Zil created nine siblings, and together the ten charmed ones rule over all. This is simply because the earliest narrators had ten fingers, nothing more. And as humans are yet to crack the language, the code, of the Universe, we continue in the tradition of using tens.

Many have come to believe there are actually only two magical beings[*]: Zil and Un. This may seem unsettling, like our entire history is a lie. In truth, it makes no difference. All that matters is that the charmed ones exist and are very *very* real. The true dispute is simply in a name. We'll give you an example. Suppose we tell you a wonderful tale all about a man named Sadik. Later, you find out that Sadik was really called Bob. Would that change things? Does that discount the wonderful adventures of Bob and his amazing powers? No. It doesn't. Making matters worse, no one would care much anyway. It turns out that Saints are the only people who wonder about Bob at all. The entire rest of the world knows about him but insists on continuing to call him by the more familiar name of Sadik anyhow.

---

[*] Cats are strong believers in this. If you aren't convinced, try telling a feline "zero zero zero one zero one." They will stop in their tracks. Look you straight in the eye. Ready for your command.

Hello! Author here. When you write a book, you aren't really thinking about the pages. You are really thinking about the words, the story, the pictures you will conjure in the heads of the readers – all that good stuff. Well, I did that. Then, when we got ready to actually print the book, we ended up with these blank pages...

Now, a blank page is a beautiful thing. It's a vast desert of space begging for content. It's a fresh canvas ready to deface with wonderful ideas, notes, thoughts, doodles, and dare we say – numbers.

When I read, I always have a pen in hand. I fill the tiny spaces to the brim – my words cling to the page's cliff, hanging on for dear life as they struggle to fit in the tiny margins the author left for me.

Thus, I had a dilemma. I could leave these pages blank so that you, the reader, would have plenty of room to fill with your own thoughts. What a benevolent author I would be! They would throw ticker-tape parades for the book who had space! But then I had a terrible nightmare. I was falling down an endless hole and all around me were the blank pages. The pages that didn't reach their full potential like little caterpillars who never emerged from their chrysalis.

Obviously, I couldn't be a part of that. I just couldn't risk it. So, throughout the book you will find pages like this. I'll try to add interesting tidbits. Sometimes mathy stuff, sometimes not. If you are a reader who is in need of space, feel free to stick a giant post-it note over me to add your own stuff. I won't be offended. And while I am filling up these blank pages – to play it safe – I'll try to leave some extra room for you as well, just in case.

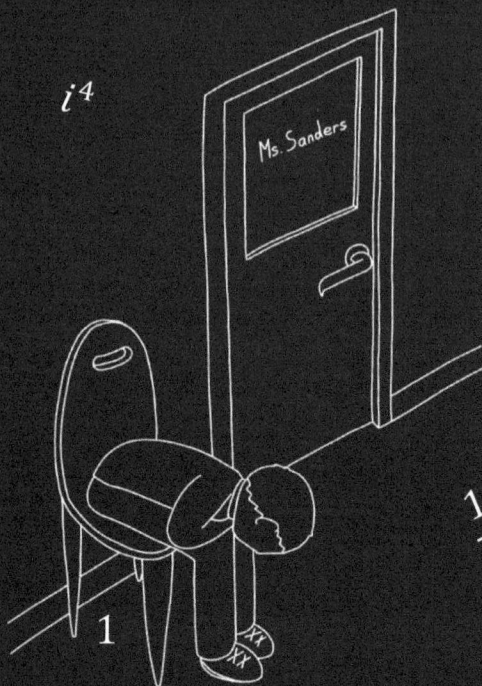

$$\dfrac{1,978,432}{1,978,432}$$

$$0.1 \times 10$$

$$i^4$$

Ms. Sanders

$$\lim_{x \to 0} \dfrac{\sin x}{x}$$

$$\dfrac{36''}{3'}$$

$$\dfrac{1}{3} \times 3$$

$$1$$

$$13 \bmod 4$$

$$2(9 - x) - (17 - 2x)$$

1

NOT A MATH KID

"SOMEHOW IT'S OKAY FOR PEOPLE TO CHUCKLE ABOUT NOT BEING GOOD AT MATH. YET, IF I SAID, 'I NEVER LEARNED TO READ', THEY'D SAY I WAS AN ILLITERATE DOLT."

-NEIL DEGRASSE TYSON

Marco was proud to say that he wasn't a math kid. It turns out, if you are lucky enough to have this label slapped on by your teachers and parents, everyone is really nice when you don't do so well on your arithmetic tests.

Marco had done okay in math for most of his life. He knew how to add and how to subtract, he had carefully memorized the multiplication tables his teachers had given him and understood the steps to multiply large numbers. Division had been more of a struggle, but he had become pretty good at those steps as well. *Put the number you are dividing under the house and the number you are dividing by outside the house. Now, how many times does the number outside the house go into the first digit (sometimes you needed the first two digits) of the number inside the house? Put that number on top of the house. Multiply. Subtract. Repeat.* Marco never understood why you put the leftovers after the letter R, but it didn't bother him much.

He knew just enough, which had always been *enough*. However, Marco wasn't in primary school anymore. He had advanced last year to secondary school and things flew downhill from there. Marco could handle the single R used in division, but recently other letters started creeping onto his homework pages. Letters like a, x, and V.

A few months ago, Marco brought home a math test that required his mother's signature. He watched as her face melted like hot cheese dripping off pizza when she read the big red letter stamped across the top of the page.

"We need to meet with your math teacher *immediately!*" his mother demanded. By 'we' she meant herself and Marco's stepfather, Peter. Marco never knew his biological dad, and Peter had been around for as long as he could remember. Despite being his only father-figure, Peter wasn't the positive influence in Marco's life that his mother liked to pretend he was.

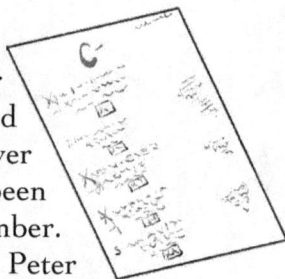

$$8 \div (5 - 3) - 4 \div 2$$

Peter didn't teach Marco to catch a ball, that was his mother. Peter didn't teach Marco to ride a bike, that was his mother, too. Marco couldn't even remember the last time Peter showed up at one of his baseball games. But, his mother, well she was always in the front row, cheering louder than anyone else. If it were just that, Marco could live with it. But it wasn't. Peter was a wicked man. It was like a new video game console that was cheaper if you bought the bundle that came with the game nobody wanted. Marco's mother was the shiny new toy and Marco was the package deal that got thrown out of sight and lost somewhere behind the couch.

Less than two days after Marco brought home the C- on his math test, he found himself sitting in a cold plastic chair, his eyes fixed on a single tile down the long, dim hallway. His ears worked overtime to pick up anything he could from the parent-teacher conference happening in the nearby classroom.

"What can we do to help him?" The shrilly whine must've come from his mother. He leaned into the door at such a sharp angle, the flimsy chair started to buckle. In the blink of an eye, Marco was on the floor. Not wanting to miss a word, the boy remained on all fours and pressed his ear to the door.

"Not everyone is a math kid. Some students just don't do well in math and that is okay. As a teacher, my job is to help my students get what they can out of the subject. We just don't have the time and resources to personally tutor every child."

Marco rolled onto his back and stared at the ceiling. He didn't need to hear anymore. He already knew what would happen next. Little bubbles began forming in the air. In the first, he saw himself at home playing his favorite video games. In the next, he was with his best friends Liam and Oliver. Another bubble displayed a scene of Marco on the baseball field, hunched over first base waiting for the ball to come hurling towards him. The last bubble had an old lady, the hump in her back was highlighted by her flowery purple dress. She pushed her thick black glasses up before

$$(4 \times 2 - 6 \div 3) \div 2$$

pulling out a wand and popping each of the other bubbles laughing as they were obliterated, vanquished as quickly as they appeared.

One of Marco's biggest gifts was his imagination. If there was a daydreaming competition, he would surely take first prize. Generally, he loved to get lost in these elaborate scenes, but not today. His mind was coming to terms with the fact that, in a few minutes, his mother would exit the classroom and demand Marco begin some after school program or meet with a tutor to improve his math scores. With only so much time in a day, he knew all his favorite activities were about to pop out of his life.

He dragged himself off the ground and plopped dramatically into the chair. To emphasize his disappointment, he slid down to maximum slouch, throwing his arms across his chest. The fact that no one was there to see his award-winning performance of 'upset kid in a hallway' didn't bother Marco. He could hear the crowd cheering as he accepted his trophy. The elaborate scene he constructed caused him to break character as he couldn't stop the grin that was inching onto his face.

The click of the classroom door opening made Marco jump to his feet. To his surprise, his mother reached out and tightly squeezed her son like she was trying to get the last drop of juice out of a lemon. Peter, without looking at Marco, lightly punched his shoulder before announcing, "Keep your chin up, little man. I wasn't a math kid either." Marco would have been annoyed with Peter, and the 'good guy' mask he always made sure to wear in public, if he wasn't so interested in this new phrase. What did it mean to be 'a math kid'? Why wasn't Marco one? Why wasn't everyone mad at him? Did this phrase somehow save him from the trouble a C- would normally bring?

Marco walked quickly down the hall, taking long strides to force his parents to keep up. He knew the after parent-teacher conference talk would happen once they were alone in the car. He remembered one year his mother had left the room with the

$$2 \times 3 + 2 - 8 \div 2$$

kindest smile on her face, only to explode the second the car doors clicked shut. This time, he couldn't wait for the talk. This time, something was different, and Marco was dying to understand why.

The trip home felt like the *It's a Small World* ride at Disneyland. Scenes whirled by as he bounced along to the friendly tune. Once his mother explained the 'not a math kid' badge that had been gifted to him, he zoomed out of the conversation and was just enjoying the journey. Here's what Marco understood:

1. Not all kids are 'math kids'.
2. For non-'math kids', like Marco, bad grades are okay.
3. There isn't really anything they can do about it. You're either a 'math kid' or you aren't.

Not only was Marco's mother no longer upset about his C-, she was beaming as she recalled his teacher's words: "Marco tries hard, he pays attention, and he always makes a strong effort on his homework." The way his mother praised him over and over for his effort reminded Marco of a puppy who just peed all over the tile floor. Sure, he's supposed to wait 'til he's outside, but he doesn't know any better and at least it wasn't all over the carpet.

The participation trophy didn't bother him. From that day on, Marco wore his 'not a math kid' badge with pride.

When the three arrived home, Marco's mother gave him another tight squeeze before sending him up to his room to finish his homework. Marco had barely sat down at his desk before he spotted the light brown ringlets bouncing up and down outside his door.

"Trouble! Trouble!" his sister giggled as she skipped into his room. Maggie was only a year younger than Marco but the combination of being the baby of the family and Peter's 'real' kid allowed her to get away with things Marco couldn't even dream of. Her attitude quickly changed as she slammed her fist on Marco's desk and shouted, "Tell me the punishment!" before swapping back to giggling again.

"I'm not in *any* trouble," Marco said proudly. "Turns out, I'm not a math kid, which means I'm *allowed* to bomb a test." He gently shoved her to the side and Maggie began to pout.

"That isn't fair!"

"The world's not fair." Marco worked hard to keep a straight face, but it was an impossible feat. The smile broke through its chains and slowly inched upwards until it invaded his face. (This was such a common occurrence that it always made him thankful he never took a liking to poker.) This time the smile-slip was worth it – he simply loved the new power he gained from his 'not a math kid' badge.

"Your math isn't even that hard," Maggie replied matter-of-factly. If Marco was 'not a math kid' then Maggie was certainly 'a math kid'. Numbers came natural to her, she could do computation acrobatics in her head, twisting and mixing the operations to suit her needs.

Marco unpacked his books and started in on his homework, practicing one of his own great skills: ignoring his little sister. Not amused, she turned and stormed out, slamming the door behind her.

❖ ❖ ❖

The next few months went pretty well for Marco. His 'not a math kid' badge was a Get Out of Jail Free card, which would've been great – if he could ever find the jail's exit. He didn't enjoy feeling completely lost in math class. So, he tried desperately to understand what was going on. He also got the feeling that his badge would only take him so far. He still had to try, to pay attention, and to do his homework for the badge to hold any power. The second time the puppy peed on the rug, he'd lose all his charm.

One of Marco's biggest problems was that math was a list of rules, and he didn't like rules. In class, he often imagined his teacher, Mrs. Sanders, was Dolores Umbridge (the strict

$$10 + (4 - 5) \times 8 \div 2$$

disciplinarian from one of Marco's favorite books, *Harry Potter*) wearing the Mad Hatter's 10/6 hat while nailing ridiculous rule after rule onto a whiteboard in his head.

One day Mrs. Sanders announced to the class, "When evaluating an expression, we always move from left to right." Knowing how important the rules were, Marco grimaced as he allowed the imaginary character to affix this decree in his head. As painful as the daydream was, Marco was relieved to see a question on his next test where he could use the rule he had forced himself to remember:

$$4 + 5 + 6 \times 2.$$

He worked left to right as instructed. First, he added $4 + 5$ and replaced the sum with a 9:

$$9 + 6 \times 2.$$

Next, he added the 9 to the 6 and scribbled a 15 in their place:

$$15 \times 2.$$

Finishing the last computation, in his neatest handwriting and larger than anything else on his paper, he wrote the number 30. He completed the problem by drawing a crisp square around his answer to ensure no one could miss it.

Marco was shocked when Mrs. Sanders returned his test the next day with an angry red X hovering over the question. He nearly fell out of his chair! He studied the paper to see his teacher

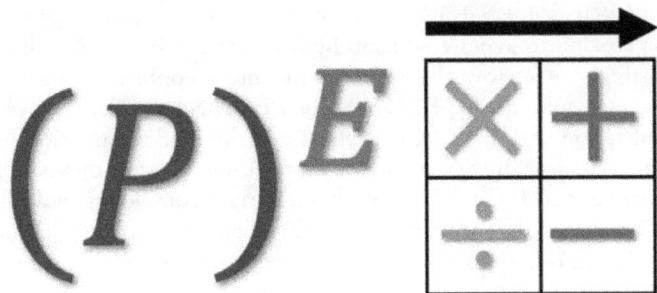

$$(P)^E \boxed{\begin{array}{c|c} \times & + \\ \hline \div & - \end{array}}$$

$$2 \times 3 + 8 \div (3 + 5)$$

had included her own answer of 21 next to the X. The 21 stuck out its tongue and laughed at Marco's confusion[*]. That day, a new rule was nailed into Marco's head: we don't *always* work from left to right. He had considered changing the first rule by crossing out the 'always' and replacing it with 'sometimes' instead of adding a new one. Since Marco wasn't totally sure when the 'sometimes' were, he thought it best to have both.

Marco continued to feel conflicted throughout the fall semester. His 'not a math kid' badge was constantly wrestling with his hatred of feeling so lost. Not knowing what else to do, he kept a running list of the rules with the hope that at least some of them would be helpful.

1. When evaluating, move from left to right

2. Don't always move from left to right, multiply first

3. Don't always multiply first, you can subtract first if there are parentheses

4. Don't trust your calculator, it lies sometimes

5. Always simplify

---

[*] In case you are wondering how Mrs. Sanders came up with 21 as the answer, here it is. The mnemonic PEMDAS stands for **P**arentheses **E**xponents **M**ultiplication **D**ivision **A**ddition **S**ubtraction which tells someone the order in which to complete the operations. As multiplication is before addition, $4 + 5 + 6 \times 2$ is evaluated as $4 + 5 + 12 = 9 + 12 = 21$. This is a completely made-up rule. It was created to avoid confusion by instituting a system we all agree to use. Interestingly, it almost always results in more confusion. Just to make things worse, it could also be PEDMAS or PEDMSA or PEMDSA because both multiplication and division, as well as addition and subtraction hold the same weight. We basically agree that we do anything in parentheses first, then evaluate any exponents. Finally, we work left to right completing multiplication or division and then left to right again finishing off with any addition or subtraction. Try it out, does the expression $6 \div 2 \times (1 + 2)$ evaluate to 1? Or to 9?

$$5 + 4 \times 3 - 18 \div 2$$

6. You can't always simplify

7. The straight parentheses mean absolute value, which is always positive

8. The absolute value problems have two answers, one positive and one negative

As he added to the list, it seemed like the new rules were always fighting with the older rules and it was impossible for all of them to coexist.

When his December report card arrived, Marco continued to earn high marks in participation making his final grade in math a solid C which made his mother happy enough. Maggie, however, threw a full-on temper-tantrum. There was no such badge for 'not a history kid' and his sister's B- earned her a week of no TV as she was 'not meeting' her teacher's expectations in participation.

It was no surprise that Marco blamed his sister for the series of unfortunate events that came next.

"I get in trouble for a B-, but *Marco*," Maggie said her brother's name like it came from her nose, "can get a C, and everyone's happy?"

"It isn't about the letter, honey," their mother replied. "You are a brilliant child. With the proper grades, you will be accepted into the best schools. The only reason your scores are lower is because you are not making an effort in class. That is something *you* can control."

"It isn't my fault my history teacher is a troll!" Maggie was stomping around, throwing her hands into the air. "He is utterly *boring*. How can you expect me to pay attention to that? I aced all my tests. He is just punishing me for knowing more than he does. Did you know that he made us memorize all the capital cities of Africa? When he told everyone that the capital of Egypt was Cairo, I quietly, *quietly*, raised my hand." She emphasized the last part, it was clearly important to her argument, though Marco wasn't sure why. Was there a way to *loudly* raise your hand? "And I told him that *actually* the Egyptians are building a new capital outside of Cairo and it didn't make much sense to memorize that Cairo was the capital since it is probably just going to change. I mean, does he even read? That is why he gave me a low classroom score. How am I supposed to control that?" In true Maggie fashion she threw herself onto the couch, arms crossed, perfect posture, and an 'I dare you to try to argue with me' stare.

"I understand Maggie-doll," their mother snuggled up next to her on the couch, so close that air wouldn't be able to find its way between them, and gently placed her hand on her daughter's leg. "Part of growing up is learning how to deal with different types of people. If anyone can figure it out, that person is you."

Clearly not satisfied, Maggie didn't make a sound. She stood straight up, then, like a cartoon soldier, turned and firmly marched out of the room. A few moments later, the echo of her bedroom door slamming rang throughout the house.

That night, Marco overheard his mother and Peter arguing. He quietly snuck down the upstairs hall hoping to catch their commentary on his sister's performance.

"He just can't keep getting C's," his mother whined. Marco grimaced and leaned in. This wasn't what he was hoping to hear, it was now something he *needed* to hear. "How will he ever get a college scholarship with these types of grades? His father was brilliant with numbers, I just don't understand."

"You heard his teacher as well as I did. He isn't a math kid. At best he's a stupid jock. His real chance is with a sports scholarship at this point. It's time you start to realize that."

"What's your problem? How could you say such a thing? Peter, I swear sometimes, it's like you don't even care. I can't leave *my* son's future up to chance. Math kid or not, he needs help. We need to get him a tutor."

$$7 + 10 \div 2 - (8 - 6)$$

The word 'tutor' left his mother's mouth and pushed through the door into the hall. Marco saw the letters *t-u-t-o-r* float closer and closer until they were right in his face, like a bully ready to start a fight. He waived them away and went back to his room. It wasn't the first and it certainly wouldn't be the last time Peter called him 'stupid'. He learned to not let that get to him. A tutor, well that was worse. That was the end of life as he knew it.

The next morning at breakfast, the air was thick with tension. Maggie was mad at everyone, their mother and Peter were clearly not seeing eye-to-eye, and for Marco, the thought of a tutor sucking away his free time had him moping around the kitchen. His mother finally broke the silence.

"Marco. Dear. We know that you are trying very hard in your math class, and we are so proud of you. But…" There it was. That one 'but' said everything. Maggie heard it, too. Her eyes lit up like a Christmas tree and she smiled, ready to relish in her brother's pain. "But," she continued, "math will only get harder next year, and we don't want you to always feel behind. We have come to the decision that we will be getting you a tutor." The fact his mother kept saying 'we' when she meant 'I' wasn't helping the situation.

Maggie looked like she had won the lottery. Marco opened his mouth to try to make his case but before he could get a word out, Peter jumped in. "We can't afford a tutor, Maryanne."

His mother had made up her mind. She stabbed her husband with her eyes, "For Marco, we *will* figure it out."

There was nothing left to say. Marco knew it wouldn't be long before Peter took his loss out on his stepson – but not in front of his wife. He'd wait 'til they were alone. For now, both boys bowed their heads pretending to focus on their plates while Maggie danced back and forth in her chair. "These eggs are *delicious*," she said with a smirk.

Marco slogged himself to bed that night feeling horrible. He weighed the situation in his head by imagining a game of table tennis. On the left, a gigantic blue paddle had the letters PROS and on the right, an identical red paddle read CONS. Marco hated not being good at math and feeling lost all the time. A tutor could help. PROS served the ball. But he wasn't a math kid, not even a tutor could fix that – that was in his genes.

CONS intercepted the serve and sent the ball flying across the table, faster than PROS could react. Marco rolled over and pulled his blanket atop his head. Even his daydreams were depressing. As he drifted off to sleep, the only thing he felt sure of was that life as he knew it was coming to an end. Everything was about to change.

In the whole history of the world, there has never been an over-the-top teenager feeling that has come true. It is the nature of the teenager to dramatize every situation, to imagine every possibility as far worse than it could ever be. However, on this night, Marco became the first. He was absolutely correct that everything was about to change. Not even Marco, with his vivid imagination and highly creative mind could possibly foresee just how devastating these changes would be.

$$5 \times 2 + 8 \div (9 - 5)$$

The story here is completely true. PEMDAS is 100% made up for the same reason someone decided mail means something we send and male means a boy or masculine. Imagine if we all agreed to, from now on, send male? E-male certainly looks cooler. *Emalé.*

Since it is completely made up, why not create our own system? Introducing SADPEM! We complete the operations in order: Subtraction, then Addition, then Division, then Parentheses, and close it out with Exponents and finally Multiplication.

In the world where SADPEM rules, what would be the value of,

$$5 - 4 \div 1 + 2 \times 9 \div 3?$$

In reality (the world of PEMDAS) is the result the same or different?

Now, maybe this helps us to understand *why* agreeing on the order is important. However, people who do math would never write an expression like this. We'd add all sorts of parentheses and brackets to make it clear what we want you to do.

$$\{5 - [4 \div (1 + 2)] \times 9\} \div 3$$

or

$$5 - [4 \div (1 + (2 \times 9)) \div 3]$$

Here's a game – write down a list of numbers with operation signs between each. Secretly decide where you want your brackets, then force your family to try different combinations until they guess the result you are looking for. If they can obtain the same result with different bracket placement, bonus points!!

$$3 + 9 \times 4 \div 2 - 8$$

$\sqrt{4}$

$\dfrac{2^4}{8}$

$\displaystyle\lim_{x \to \infty} \dfrac{1}{x} + 2$

$\left(\dfrac{\frac{6}{4}}{3}\right)^{-1}$

$2!$

$2 \times 2 \div 2 + 0$

$f(1): f(x) = (x+2)^2 - 7$

2

THE TUTOR

"MATHEMATICS COMPARES
THE MOST DIVERSE
PHENOMENA AND
DISCOVERS THE SECRET
ANALOGIES THAT UNITE
THEM."

-JOSEPH FOURIER

T he most sought-after time of year had arrived – winter break. Making things even better, Marco had heard nothing more about his tutor and no news was good news. He spent his days with his best friends Liam and Oliver. The boys had very competitive snowball fights in the park which turned into very competitive video game tournaments inside after the sun had disappeared behind the clouds or the horizon, whichever came first.

Maggie and Marco were even getting along. They tore open their Christmas presents and spent the entire week after the holiday playing this game or that. The family sipped hot chocolate and watched holiday movies after dinners. His mother had received a real popcorn machine from Peter as a present and the freshly exploded kernels dripping with warm butter and dusted in salt tasted better than the theater's.

January came more quickly than expected. A sickening feeling of dread began to fester within Marco. He was not at all excited to return to school. The news his mother shared over breakfast only made things worse.

"I was speaking to Jan the other day. You know Jan… she works at the library." She flipped the pancake on the stove before turning to face her family who were all gathered around the kitchen table with heads buried in a device. Peter was swiping through the morning news stories on his phone while Maggie and Marco were playing a game of air hockey on the refurbished iPad they had received as a joint Christmas gift. No one was really sure who she was talking to, so they all half-listened.

She turned back to the stove and plopped a large spoonful of batter into the pan. "Well, it turns out she knows this man who used to teach math at the middle school two towns over. Anyway, apparently, he has retired. And you know how it is, he's getting lonely and looking for work." Neither Maggie, Marco, nor Peter had any idea what it was like to retire. Thus, they all naturally

$$\frac{\$30}{60} = \frac{x}{32}$$

assumed she wasn't talking to them and continued what they were doing.

"Here's the best part!" Maggie and Marco looked up to see their mother's arms vibrating with excitement. A bit of pancake batter flew off the spatula she was holding, *splat* onto the door of the refrigerator. "Jan reached out to him, and he said he would be happy to tutor Marco!" Now everyone was paying attention. "AND…" Marco's mother looked directly at Peter as if to say, 'this next part's for you'. "He is only asking for $30 a week! Can you imagine? It's like it's meant to be."

Peter, clearly still angry about the situation, gave a slight nod and continued fiddling with his phone in an unsuccessful attempt to disguise his feelings. Maggie was thrilled. She did a little na-na-na-na-boo-boo shake in Marco's direction. The winter break bliss was officially over. Marco expected the news to hit him harder than it did but he had seen this coming for weeks which had given him time to accept his fate. His mother went on to say that his tutor, Mr. Pikake, would meet Marco in the school library after last bell.

On their way out, Maggie whispered, "Goodbye to life! Goodbye, goodbye!" in her best Bette Midler[*] impersonation trying desperately to get on her brother's nerves.

The day zoomed by. Before he knew it, Marco found himself staring down the long hallway to the library struggling to make his limbs move. He slowly dragged himself, one step at a time, until he eventually reached the tall wooden double doors. Giving them a shove, he peered inside. Empty.

Hopeful the conversation at breakfast had been a bad dream, Marco found a table, dropped his backpack to the ground, and slouched down low in the chair. He pulled his hoodie over his

---

[*] It is highly likely that if you are reading this you have no idea who Bette Midler is so we will help you out. Maggie is referencing one of her favorite movies *Hocus Pocus*.

$$\frac{\$30}{85} = \frac{6}{x}$$

thick black hair in an attempt to conceal his identity and began to discreetly examine the room. A large platform that resembled a judge's bench sat next to the doors. Atop it, an older lady was perched reading a tome that looked thick enough to be the dictionary. Lines of bookcases surrounded the two rows of four-top tables in the center of the room. Marco saw a few students searching for books in the stacks, another two sat at the table farthest from him – still no sign of his tutor. He bent down, pulled his math book from his backpack and opened it on the table in front of him. Pretending to read was another attempt at disguise. He hoped that if his tutor did show up, he would overlook Marco as he didn't appear to be waiting for anyone.

The sudden squeeze of a firm hand followed by a booming voice caused Marco to jump. "Marco?" He turned to see a very tall, very thin man looking down on him. The man had short salt-and-pepper hair neatly trimmed but tussled as though it hadn't been combed in a day or two. He wore a light blue colored shirt, a forest green vest, and a grey suit jacket. All three clashed with the orange and red striped tie that hung from his neck. A smile slowly formed on the man's face that appeared to unnaturally reach from one ear to the other.

"I'm Mr. Pikake, you can call me professor, I believe we will be working together?" Each word flew out of his mouth like a bullet. The tutor displayed perfect pronunciation that resulted in a sharpness of speech Marco had not heard before.

The student was frozen. His senses and his brain were overwhelmed. He imagined pushing back his chair and darting out of the double doors – they couldn't have been more than thirty feet away. The scene that played out in his head however, ended up with both Mr. Pikake and Marco on the floor. When Marco sprung up, he accidently pushed the tutor backwards before tripping on his own backpack in his attempt to flee. Deciding that neither running nor hiding was a realistic option, Marco settled on a nod.

$$\frac{\$30}{x} = \frac{5}{3}$$

"Wonderful!" Mr. Pikake's voice echoed. The librarian shot him a look that unmistakably said, 'shut up'. Marco was confused. He had an excellent memory and could have sworn his mother had mentioned Mr. Pikake and the librarian were friends. His eyes scanned up the judge's bench to the gold-plated sign that read *Sabrina Samaritan, Head Librarian.* It was then that Marco realized there was more than one librarian in the world, which was something he had surprisingly never considered before.

In only two quick steps, the man carried himself to the other side of the table and took a seat across from Marco. "This is a wonderful school year for math!" Mr. Pikake continued in a loud whisper. He looked over his shoulder and winked at Head Librarian Sabrina, who blushed. "I spoke with your parents and your teacher about how you have been doing. I heard you are an expert in the basic operations." He held up his hand to reveal four slender bony fingers. He pointed to each as he recited, "Addition, subtraction, multiplication, and division." This was followed by a chuckle that made Marco uncomfortable. The man appeared to be some strange hybrid of the elderly and a kindergartener. "So, tell me. I have heard from everyone *but* you. How do you feel about math?"

Marco was surprised by both Mr. Pikake's words and his sentiment. It was clear that the tutor was genuinely interested in *Marco's* thoughts, not only what everyone else had said. This had the effect of building an instant, but unspoken, trust between the two.

"Well. *Ummm.*" Marco started. "I actually used to like math. It was all about counting really, just in different ways." As he continued, a wave of calm washed over him. The fight-or-flight reaction was subsiding with each new word. "I even found math to be really helpful. I used it all the time in this video game I play. Like a sword has strength 18 and an axe has strength 12. If I came across a zombie with 40 health, I knew it would need three good hits with my sword or four good hits with my axe to beat it. And I'd have to do quick math, too. I almost *never* have tools with full

$$\frac{38}{\$30} = \frac{x}{15}$$

strength. I maybe have only two good hits left in my sword, so I'd have to make a decision fast." Marco raised his head and for the first-time locked eyes with Mr. Pikake. He found a warmness in them that broadcasted the tutor was both listening and interested. "But then, math got a lot harder. Now it's all about rules and letters, and I don't like either."

Marco bowed his head returning his gaze to a small blemish on the library table. He stared at it so hard he thought for a moment he could magically fix it with his mind. His face twisted into a squint as he harnessed all his brainpower on the dot. *Why is he not talking?* The silence made Marco both uncomfortable and annoyed. Finally, Mr. Pikake spoke up. "Riveting reflection. I think you will be happily hypnotized once we start working together. You see, your zombie math *is* math with letters!" Marco smiled as he imagined the zombies doing the math. "You have talentedly trained yourself to think about things differently, which is a stupendous superpower. In math, we are the gods, the creators. We dominate and together we will harness and extend your power."

Marco had never heard anyone describe math this way. As weird as it was, he didn't care because all he could focus on was the giant balloon inflating in his chest that forced him to sit up straighter and made his chin raise at least two inches. Mr. Pikake's words made Marco feel like a superhero, and he loved it.

"We will get started tomorrow. Commencing with counting." Mr. Pikake snapped Marco back to reality. He could hear the hissing of the balloon losing air as he returned to standard slouch.

"B-B-But..." Marco stuttered, still recovering, "I know how to count!"

"*Ah.* Precisely why it is the perfect place to begin." With that, the tutor pushed back his chair and sprung to his feet. Marco had forgotten how tall the man was and felt tiny beneath his commanding presence. Mr. Pikake pivoted on one foot and began walking towards the double doors. His long arms and legs swung

$$\frac{x}{2} = \frac{\$30}{3}$$

effortlessly, reminding Marco of an Enderman from *Minecraft*. Before he left, Mr. Pikake gave a kind nod to the librarian, who blushed again. In the blink of an eye, he was gone as if teleported to a new world.

Marco gathered his things. He was surprised at how good he felt. Maybe tutoring wouldn't be as horrible as he had imagined. If Mr. Pikake could help him, he'd certainly be happier, his parents would be happier, and his teacher would probably be happier, too.

The tallest of the six drummers in the school band, Marco's best friend, Liam also had an after-school commitment. He sat waiting on the bench outside the library where they'd agreed to meet. His light brown hair tussled as if he'd been head banging for the last hour. The second Marco appeared, Liam sprung to his feet, "*Dude.* How'd it go?"

The two laughed as they walked down the hallway and Marco described the encounter. "He is basically an Enderman who dominates numbers instead of blocks. I'm gonna steal this 6 and put it over there," Marco joked. The boys could barely stand, they were overcome with hilarity.

Liam placed his hand on Marco's shoulder, "He's probably Illuminati. They love to get into kids' brains. Just be careful, dude." Not only did Liam have a skill for killing a good moment, there was never a conversation he couldn't find a way to slip a conspiracy theory into. Marco gave his friend a playful shove out the doors.

Two trucks sat in front of them, waiting. Liam shot Marco a sympathetic look before racing to the shiny blue Tundra leaving his friend alone to slowly inch towards the beat-up Ford. Peter's truck resembled its driver. Short and stout, it shouted to the world 'I don't care what you think'. On the passenger door, an image of a gigantic cockroach was painted along with the words *The Terminator*. Peter worked in pest control which always seemed backwards to Marco who thought his stepfather was the real

$$\frac{x}{70} = \frac{9}{\$30}$$

insect. Marco always had the feeling that, if he could, Peter would jump at the chance to exterminate him. As the January weather attacked like ice arrows being shot from invisible fairies, he considered the assault a pre-match for what was about to come.

"You're such a little shit," Peter mumbled as Marco opened the truck's door. "Your buddy ran over, but not you. You just take your time like the whole world will wait for you." Knowing better than to say anything, Marco slammed the door and buckled his seat belt. "I'm gonna tell you one thing and I'm only gonna say it once. So you better be listening. I ain't got the time or the money to waste on a thick head like yours. Your mother's got it in her mind that this tutor can help you, but we both know better than that. You were born stupid." Peter turned to look at Marco who was staring out the window. He shoved the boy causing the truck to swerve sharply. "You listening to me?" Marco only nodded. "I'm not gonna pay to help the helpless. Might as well throw my money right into the trash can. You're gonna tell your mother this isn't gonna work. Make it up for all I care. Tell her the tutor's mean, tell her he hit you, tell her whatever. But this nonsense better stop, and it better stop soon."

Marco's blood began to boil. He thought about Mr. Pikake's words, 'dominate' and 'power'. He imagined himself standing tall, like Superman, his cape waved in the wind behind him and with a flick of his wrist he sent numbers flying here and there. The idea that Marco could control the math rather than the math controlling him was liberating. The idea that Marco could control his own life rather than live under Peter's thumb was desirable. He needed a way out and he saw Mr. Pikake as that way. While Peter made Marco feel small, Mr. Pikake had built him up into something strong. With all his might Marco held onto that feeling, he bit down hard on his lip until the sweet metallic taste of blood filled his mouth. He focused on the taste to drown out Peter the whole ride home.

When Marco entered his room, the sight of Maggie cheerily bouncing on his bed infuriated him. "I've been waiting

$$\frac{55}{x} = \frac{\$30}{12}$$

FOREVER!" she jumped again to her knees before landing on her stomach and placing her hands below her chin like a baby angel posing for a picture. "Did your tutor quit after seeing you're a lost cause?" she joked. Maggie had inherited not only her father's sharp tongue but also his eyes...deep brown orbs that seemed to refuse to allow light to penetrate them.

Seeing Peter in the body of a defenseless child sent an overwhelming urge to rip her apart down his spine. Without even a second to think, Marco pounced. As he flew towards his sister, her kind and loving smile filled his vision. It was too late. Gravity was in control now. At the last moment, he managed to twist his torso just enough to change his trajectory – with a loud thump he landed next to her on the bed. He wrapped his arms around his little sister and started tickling. Maggie shrieked in delight, her arms not quite long enough to retaliate.

A new feeling swelled inside Marco. He felt powerful. He never ever wanted to let this sensation go. It was a risk to keep up the tutoring, he'd have to deal with Peter's wrath. What Mr. Pikake had described, *'we are the gods, we dominate'*, was something Marco desperately needed. He yearned to control something, anything. As he looked around the room, for the first time he noticed just how many numbers were near him, in the shadows, inconspicuous. They were on his clothes, his clock, on every bag of chips and bottle of soda. *Controlling you will do just fine,* Marco thought. As he gazed upon the numbers, he didn't notice that they were also looking back at him.

$$\frac{69}{\$30} = \frac{x}{10}$$

3

IN THE BEGINNING

"GO DOWN DEEP ENOUGH
INTO ANYTHING AND YOU
WILL FIND MATHEMATICS."

-DEAN SCHLICTER

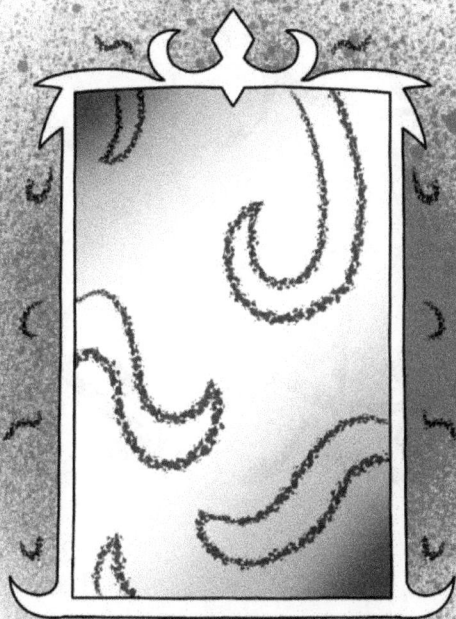

One of the few things that existed before the Universe was created were Spogs. Enchanted particles, Zil used the Spogs to form his nine siblings – the charmed figures. Like constellations, the arrangement of the Spogs dictated their shape.

Un   Zwei   Arbah

Shi   Kween   Exee

Septem   Acht   Nava

Only nine of the ten Spog figures are known. The tenth, Zil, is somewhat of a mystery. It has been described as a ghost – something that is there, but also isn't. Many believe the more Spogs, the stronger the charmed one, but strength can be deceiving. For each sibling possessed unique and interesting traits. Figure ten, the absence of Spogs, is one of the most powerful formations.

The siblings saw Un as little and weak – made up of only a single Spog. What they didn't know, was that Un's image would be extraordinary. Not only would she be the foundation for everything to come, she also possessed the power of reflection, gifted to her by the Mirror of Wonders.

Zwei was the symbol of harmony, bringing peace everywhere he travelled. For everything made from Zwei would have a natural balance – the ability to split perfectly into two equal forms.

Arbah was filled with creativity. Talented, she could bring visions to life.

Shi was known for his stability and was often called upon to build, while Kween was charmed with change – greatly feared and respected by all.

Exee was known as the perfect charmed one and Septem was filled with secrets – having the power to unlock knowledge the Mirror of Wonders presented.

Acht brought about order and Nava, well, it took Nava some time to understand herself.

THE FABLE OF NAVA

When Nava emerged, all that existed was dust, darkness, and her siblings. For many years, Nava was regarded as the most powerful amongst her family simply because she was made from more Spogs than any of her brothers or sisters. However, they soon learned that Spogs were not the only thing making Nava special, she possessed other remarkable traits as well.

The charmed ones crafted from Zil all possessed a spellbinding power called fracturing. Like fireworks, they could tear themselves apart into an amazing explosion of singular Spogs. But for the daughters of the Universe, aside from fracturing, they had an unbreakable form. The brothers, Zwei, Shi, Exee, and Acht all had the ability to split their Spogs in perfect halves while the daughters, Un, Arbah, Kween, Septem, and Nava were given a special crown – a Spog that shone atop their head and locked in their shape. The crown was both powerful and limiting. For the daughters were resilient, they could not be torn, but they were bound to their shape for eternity, never otherwise able to transform.

Whether by fate or by design, unlike her sisters, Nava was the sole daughter of the universe who was royal, yet malleable. While she did not have the power to split into halves like her brothers, she shared with Exee the mighty trinity – the ability to be torn into three equal forms.

Nava's unique traits led to isolation. Her fear of being alone quickly turned to anger. "They deny me from being a brother because of the crown that sits atop my head. They deny me from being a sister because I am more flexible than them, able to split. They all fear me for my Spogs are vast." In that moment, Nava decided that while she was stronger than her siblings, it wasn't enough. In search of more power, Nava travelled to the Mirror of Wonders.

The Mirror of Wonders lay on the far edge of time, it was said to hold the secrets of space. As Nava peered into the mirror, she asked, "I wish to be the most powerful of all time. I wish to make my brothers and sisters bend before me and accept my greatness. How can I accomplish this, Mirror?"

The Mirror of Wonders churned to life. Red, purple, blue, and yellow dust swirled across its face as it searched to locate the answer Nava sought. Nava listened carefully as the Mirror replied.

Daughter Nava, powerful you are yet you stand here alone.

The influence you seek would require a clone.

The charmed ones gain strength through links in a chain.

The position yields more power than a name.

You are who you are, that cannot be changed.

With your siblings, together you can rearrange.

The Mirror showed Nava a long chain. She sat in the first position, ordinary, herself. Then Nava began to move down the chain. As she travelled to the left each link brought her more power, she was bigger, stronger. Nava understood. Being the first link in the chain would never give her the influence she desired. She was only one Spog more than Acht, not enough to make much of a difference, but with a chain – then she could rule over her siblings. With that authority, they would no longer be able to deny her.

Making quick work, Nava used the magic of her Spogs to gather the surrounding dust and form them into two links. She attached the links to her side and admired herself in the Mirror. Before Nava began her journey home, the Mirror left her with a final warning.

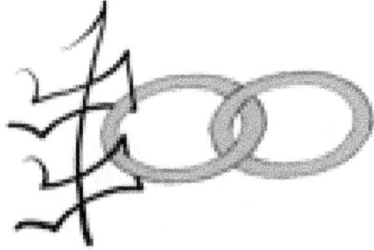

You must gain followers to willingly stand by your side.

Alone, you are nothing more than filled with pride.

As Nava made her way back to her siblings, she felt satisfied as her two chains followed along, but also worried. She knew only Septem could truly understand the messages from the Mirror. Pushing the thought away, she grew only more determined. Why was Septem so special? She had no problem understanding the Mirror's words on her own.

When she returned, Nava tortured her siblings. She explained to them the power of her chains and with it, forced them to follow her every command. The day quickly arrived when Un refused to continue under Nava's tyrannical rule. "I don't care how powerful you are, Nava. I will no longer obey you." Surprisingly, Zwei stood next to his sister and agreed. "I stand with Un." Zwei declared, "Your rule is no more!"

Infuriated, Nava demanded they travel to the Great Scale. As Un and Zwei already possessed the fewest Spogs, Nava knew it would overwhelmingly rule in favor of her and her links forcing her siblings to fall in line. The three set off on their journey. Scared, Un and Zwei huddled together throughout the trip. When they arrived, Nava pushed her siblings onto the scale and was shocked by what she saw. A magical link glimmered, attaching them at the hip[*].

Nava didn't care. The scales would still declare her as the most powerful, she had two links and more Spogs. She climbed onto the platform. The scale waivered back and forth, back and forth, before finally declaring Un and Zwei as the victors.

The figures were baffled, it seemed impossible that Un and Zwei could ever hold more power than Nava, especially with her links. As Nava recalled the Mirror's words, she burst into tears. She realized the links she had fashioned held no power. Without a sibling at her side, she was only Nava. It was Zwei's willingness to stand with Un that created a true chain.

The family was quick to forgive Nava and her transgressions. She shared with them the words from the Mirror of Wonders, and it was then that the siblings finally realized their true power. Septem explained that when they stood together, they amplified their strength. When Zwei followed Un, Un became much more than she was alone. It forced her into the second position on the chain which – since there were ten siblings total – represented ten Spogs, more than any single charmed one possessed individually. With Zwei by her side, together they harnessed the power of twelve Spogs – more than they ever thought possible. The charmed ones spent the next years

---

[*] Today we recognize the link created with Zwei following Un as 12.

experimenting and perfecting their abilities. They found that if Un followed Zwei, they were stronger than if Zwei followed Un. They found that more cooperation led to greater results. If Zwei and Arbah followed Un, they surpassed Septum following Acht. And perhaps most importantly, they found that they were always strongest with Nava in the lead[*].

With their newfound knowledge, Arbah used her powers to bring their visions to life while Shi built everything they could possibly dream up. Together, the charmed ones began to construct the Universe and it all started with Numberville.

* * *

The Fable of Nava produced a wealth of important discoveries. As with most fables, at its heart the story speaks to the dangers of pride, a lust for power, and the beauty and strength of teamwork. More crucially it provides us with a deep understanding of the origins and ancestry of Numberfolk.

Just beginning to experiment with their powers, the ten charmed figures created the first Numberfolk in their own image. They named this inaugural group of children the *digits* and we recognize them today as $0, 1, 2, 3, 4, 5, 6, 7, 8,$ and 9. Next, they fashioned links placing combinations of digits together to construct $10, 11,$ and 12. They constructed 100 and $1,000$. They constructed $1,467,892$ and they constructed $42,899,456,612$ and they constructed everything in between and everything after.

To make a comparison to humans, all living things are determined by DNA. There are four building blocks of DNA – four digits – and while extensive, a DNA strand has a finite length of about 6 feet. Now consider Numberfolk. They have ten building blocks at their disposal, far out numbering the human's four. Worse, Numberfolk have no limit

---

[*] We understand this as $21 > 12$, $123 > 87$, and of course the largest n-digit number has a 9 in every place.

on how long their chains may be*. That is to say, Numberfolk surpass the human ability. There is no ending on how many Numberfolk exist, they continue forever and ever.

Infinity was the first problem the charmed ones had to overcome. Shi raised infinite homes for their children and as each Numberfolk was assigned to a structure, the village quickly became full. Unfortunately, the charmed ones were nowhere near completion – they needed to solve their vacancy problem. Zwei, wanting to end the conflict, was the first to suggest a solution, "Since there is no end to the village, we can simply find the last resident and continue from there."

The charmed ones searched and searched for the final resident, but their task was impossible. Like Sisyphus pushing the boulder up a hill, each time they got close, they found themselves again at the beginning. Because the village was infinite, there was no final resident to find, for each time the charmed ones believed to have found the final resident, there was always another which was one more. Their creation was both never ending and completely occupied.

It was Exee in their perfection that ultimately found the solution to their conundrum. "There is no end, but there is certainly a beginning. We need not find the final Numberfolk, we can simply instruct each resident to move one house down. The first home will then be empty for us to create another."

And so it was done. The charmed ones continued to create and create. Before disappearing into the ether to allow their sons and daughters to live out their lives unobstructed, their final act was to christen their creation with a name – Natural. To the charmed ones and their offspring, all was good.

---

* If you are having trouble believing this, test it out. If you don't have paper that is over 6 feet in length, try using sidewalk chalk. Start writing digits until your number chain is longer than your DNA. Just like the building blocks of DNA, digits can appear more than once in a chain.

Little did they know, the Great Scale had a different idea. Being omniscient, the Great Scale understood that everything in the Universe required balance. Ordering the Mirror of Wonders to spy on the charmed ones, for every being they created, the Mirror created another – a twin. One might even say in the interest of balance, if one was good then the other, well, they must be evil.

$4 + (4 - 4) \times 4$

$\int_0^{\sqrt{8}} x\,dx$

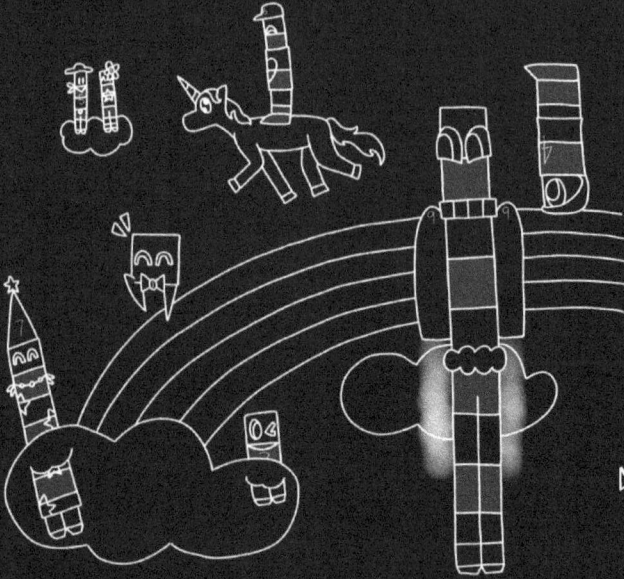

$\dfrac{4!}{3!}$

2^2

$(2 - x)(2 + x) . x = 0$

$\sqrt[3]{64}$

4

COUNTING

*"WE LEARNED TO BE HAPPY.
WE DANCED 'ROUND THE
HALL. AND LEARNING TO
COUNT WAS THE KEY TO IT
ALL."*

-SESAME STREET

W hen the final bell rang on Tuesday, Marco rushed to the library. He lit up when he saw Mr. Pikake already sitting at the same table where they had performed their initial introductions.

"Perfectly punctual!" Mr. Pikake greeted Marco with his Cheshire cat grin. "Are we ready to begin?" Marco nodded, pushing down the feeling of fear that was creeping into his throat. Being a teenager had been like riding a rollercoaster recently. He could go from feeling wonderfully happy to devastatingly sad in an instant. His mother had attributed the dips and dives to normal hormones. Whatever it was, Marco hoped it would stop soon, it was beginning to make him nauseous. He realized, that in his excitement about his quirky new tutor, he had set his expectations so high they might be impossible to meet. This was after all *only* a math lesson. But the power the professor had hinted at when they met – well, that was too much to turn down. Marco needed that power. Forcing the feeling away, he sat across from his tutor and pulled out a notebook and pencil.

"You won't need those!" Mr. Pikake boomed. Quickly catching himself, he glanced at Head Librarian Sabrina. Surprisingly, she smiled in return. He bent down and picked up his briefcase, a soft brown leather messenger bag that surpassed him in age. Reaching his long fingers into the bag, he revealed a jumbo pack of M&M's. He scooped a large cupful from the pack and dumped the chocolates on the table.

"How many are there?" Mr. Pikake asked the question as if expecting an immediate answer. Marco was instantly reminded of his 'not a math kid' badge. *We haven't even started, and I am already failing*, he thought to himself. The question reminded Marco of the carnival games that displayed a large glass jar of candy and asked children to determine how many were in the jar. There must've been some secret trick Marco missed, as he was always hundreds away from the correct number and had never managed to win the jar. His sister, on the other hand, never failed to pull out a reasonable estimate or to remind Marco of her superiority as she gleefully chomped down her prizes in front of him. He shuddered as he considered the possibility that Peter was right. He couldn't let Peter be right.

"How could I possibly know that?" Marco responded sharply, revealing more of his emotion than he had intended.

"By counting them of course!" The tutor either missed or ignored Marco's negativity as he remained as cheerful as ever.

The two began counting the M&M's. Marco constructed piles of tens which caught Mr. Pikake's eye. "Interesting method you have there... explain."

"It's too hard to keep track counting one-by-one. I'm making piles of ten, so I can just count the piles, 10, 20, 30, 40, whatever."

"Perfectly to the point. *How* we count can be powerful."

Marco perked up, maybe this wasn't as hard as he'd imagined. Afterall, he had been counting for as long as he could remember, it was certainly something he could do well.

The two continued making piles of ten candies until everything was grouped. Marco looked over the table and didn't need to count the piles to know how many M&M's the tutor had dumped in front of him. Each pile was a small circular mound that together resembled the face of a familiar die. He arranged the piles into two lines of three. Next to the mounds lay the leftovers. He could easily identify there were two groups of four.

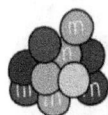

"How many are there?" Mr. Pikake restated the original question as if asking for the first time.

"68."

"Is that all?"

Marco thought he was sure but instantly began to doubt himself. Had a rogue chocolate rolled out of sight? Did one of his mounds miss the mark as a perfect ten? "Yesss…" he replied. His 's' lingered making his response more of a question than a statement.

"You are both correct *and* mistaken!" Mr. Pikake chirped excitedly.

The sharp pull of gravity was jarring as Marco's internal rollercoaster plunged. He imagined himself in the leading car, his 'not a math kid' badge tattooed on his forehead. Frustration took over and he snapped, "There's either 68 or there isn't. That's one of the good things about math, there's *always* only one answer."

"*Ahh*. I see the brazen blocker! You are correct that there are 68 candies. But," the last word popped from the tutor's mouth, "numbers are more like people than you know. You are Marco, you will always be Marco, but," another pop, "you are not *only* Marco."

Numbers are like people? Marco had sprouted ears and was tumbling down the rabbit hole. As crazy as it was, he was hypnotized by the idea. He understood people much better than numbers. If they really are alike, he just might be able to master math after all.

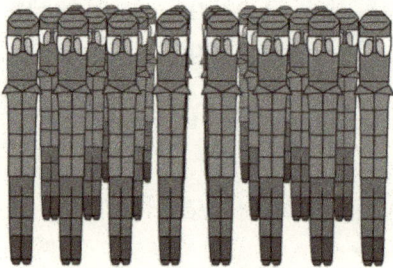

"Let me explain," Mr. Pikake leaned in. "You had the marvelous mind to count by tens. Some days we feel like tens. Groups of ten are strong, unwavering, they are the basis of the number system!" Marco envisioned lines of soldiers. Each an identical copy of their neighbor, constructed of blocks standing ten high. They marched together in step as Mr. Pikake perched above them waving his hands as their conductor.

"Yet, we are not always strong. There is a rainbow of emotions we experience!" As Mr. Pikake continued, so did Marco's imagination. The ten block soldiers burst at the seams to become 1 and 9, or 2 and 8, and every other possible combination. They frolicked happily across a rainbow, some skipping, others riding unicorns, some even had jetpacks that spurt out colors rather than fire.

"As humans, we often find ourselves split. Have you ever felt torn, Marco?"

The idyllic rainbow-unicorn world divided like ripping paper as Marco searched his memory. It didn't take long to find the first time he had ever truly felt stuck in the middle – it was Halloween of his 5th grade year. He had always gone trick-or-treating with his best friends Liam and Oliver since they first met in their kindergarten classroom.

That year, Liam spent days planning their perfect route. He made sure it passed by Mr. Markle's house, who always gave the boys two giant handfuls of the best kinds of chocolate, and by the Lucas' home who gifted each visitor with a handmade fortune cookie. The boys would open their fortunes together before the end of the night. When they were little, they believed the Lucas' had supernatural powers, able to foresee the year to come. The tradition continued as they aged, and they were always eager to unwrap their future.

It was Mrs. Lucas who prompted their tiny imaginations to run wild. The boys would ring the bell and she would appear as a silhouette in the doorway so quickly she must've had unnatural foresight. Carrying a large silver bowl twice her size, she never blindly grabbed a cookie from the cauldron, but lovingly peered at each one, as a mother gazes down on her children, before carefully selecting the perfect cookie for each boy.

One year, Oliver's fortune read, "Tough times ahead. Things will be alright. You are loved both day and night." The three friends went wild guessing what the cryptic words might mean. The shock came two months later, right after Christmas, when Oliver's parents announced they would be getting a divorce.

Maybe it was the fortunes that caused Oliver to declare he would not be trick-or-treating that year. He wanted to abandon their tradition in favor of the school's Fall Festival. The argument he made was that not only were they getting too old for trick-or-treating, but it was their last year at primary school, and he wanted to explore the haunted house, contests, and games they had to offer.

The choice fell to Marco. No matter what he picked, he would be disappointing someone, and the decision felt impossible. He ultimately decided to go trick-or-treating – it was their tradition and Liam had worked so hard. Oliver dragged himself along, but it was months before their friendship felt back to normal.

Marco's mind travelled back to Mrs. Lucas' fortunes. He struggled to remember what his read for this year. He was rummaging around his brain trying to dig it out when he noticed Mr. Pikake's stare. Having taken far too long to answer the question, he had forgotten what had even been asked.

"Oh...*umm*...What was the question again?" A hot embarrassment rose up his face.

Mr. Pikake chuckled, "Based on your expression, I am guessing you have felt torn before?"

"Yes. Yes." Marco started remembering where they left off, "I was thinking about how I was torn between two friends."

"Wonderful!" Mr. Pikake realized his excitement was uncalled for and corrected himself, "I mean, horrible thing to be torn, though you can understand how a Numberfolk like 68 could be torn, too?" For anyone else, Mr. Pikake's use of the term

'Numberfolk' to describe quantities would be absurd. For Marco, it only fueled his imagination.

The tutor's long arm swept across the table destroying Marco's perfect ten mounds to reveal a divide between two new piles. "Now 68 is 23 + 45."

Marco finally saw what the man was trying to demonstrate. "I get it. You are saying that numbers, like people, are not one thing. They can change depending on the situation." The thought of Peter's many masks shot through him like the icy sensation of a cold drink on a hot day and made Marco more determined than ever to follow along.

"Exactly, my boy! Numbers are chameleons. Horribly tricky little things. They disguise themselves into many different shapes and forms. For 68 could be four piles of 17 or thirteen piles of 5 with 3 leftover." Marco was transfixed as Mr. Pikake, like a magician, waived his hand and began shuffling the M&M's about the table, each time making a new combination.

"68 could be 25 + 43," Marco chimed in.

"Or 34 × 2," the tutor echoed back.

"How about 70 − 2?"

"Yes! And 6 × 11 plus 2 more!"

"And 67 + 1!" Marco was enjoying this game. Not only was it easy, he was beginning to feel the power the professor had suggested. He could make 68 into whatever he wanted it to be, within certain limitations.

"Or one million, six hundred and seventy-eight thousand, two hundred and fourteen minus one million, six hundred and seventy-eight thousand, one hundred and forty-six!" Mr. Pikake let out a jolly Santa Claus chuckle. "There is no limit to the ways Numberfolk can disguise themselves!"

Just then, the professor slammed his hands on the table and leaned in close to Marco. In a low voice he continued, "I will share with you a secret, Marco. The more familiar you are with the many disguises, the more control you will have. Numberfolk are simple beings, but we, we have the power to tear them apart and stitch them back together. We can force them to do our bidding, and we can find them no matter how impressive the mask they wear."

As Marco looked into Mr. Pikake's eyes, he was reminded of a vocabulary word from his English class – *maniacal*. He had thought it was a word he would never have cause to use, but the way the old man's irises glimmered like a raging, hot blue fire as he spoke, Marco could think of no better-fitting description.

"This brings us to my primary pontification!" the professor shouted throwing up his hand. A loud '*sshhhhhhusssshhh*' came from Head Librarian Sabrina as Marco tried to translate what the tutor had even said. "Let us begin with the holy Numberfolk, the ones we can count." Mr. Pikake raised each of his lengthy digits as he said, "one, two, three, four, five." Marco started to catch on. The professor was saying they were going to talk about the counting numbers, which was good news since Marco hated fractions.

"Know your enemy!" he continued in a quiet yell as to obey the rules of the room. "We can classify these Numberfolk in many ways such as even or odd."

"I know about that!" Marco was happy the lesson wasn't getting too hard. "Even numbers are the ones you get when counting by twos. Odd numbers are all the rest."

"Perfection, boy! But…" he popped, "we are interested in a new category of classification. We wish to group by prime or by composite." The rollercoaster was picking up speed. Marco had heard of the word 'prime' but wasn't entirely sure what it meant. "The counting Numberfolk are composed of clusters." Marco's ride was going faster now, a sickly feeling started to form in his stomach. "Like LEGO." The rollercoaster came to a sharp stop.

Marco knew all about LEGO, he leaned in as the professor continued. "Imagine you have blocks that are two tall and blocks that are three tall. You don't have any blocks that are four tall – why would you need them?! You could just use your two-blocks to make a four. You have blocks that are five tall as well." He turned to Marco and asked, "Do you need six tall blocks?"

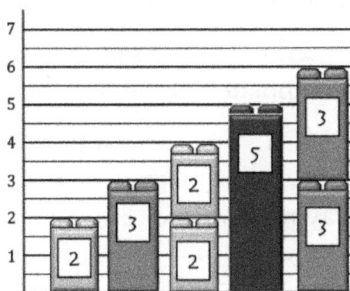

Considering the question, Marco thought about some of his craziest builds. He was constructing a castle, the turrets needed to be six blocks tall, he searched his gear – no six. He grabbed three two-blocks but didn't like the way they looked. Scanning his material again he saw three-blocks, using two of those he completed the turrets. A smile swelled across his face. "No, we don't need six-blocks. We can use twos to make six or we can use three's instead!"

"Splendid! You see, all counting Numberfolk are composed of these clusters of blocks. How would you build a thirteen?"

Without thinking Marco blurted out, "I'd use two six-blocks and one more!" As the words flew from his mouth he realized his mistake, "Oh, I don't have six-blocks. I'd use four three-blocks and one more, I mean." He shuffled uncomfortably in his chair.

"*Ah*, there's the rub. You may not use different types of blocks in your build! If you choose to construct with three-blocks you may only use three-blocks."

This made the question much harder. Marco couldn't use two-blocks to build thirteen, because thirteen was odd – it wasn't made up of twos. He couldn't use three-blocks either, or five-blocks, or seven-blocks. "I don't think it is possible," Marco muttered.

"Very good! A thirteen can only be made using one-blocks, or if you are lucky enough to have a thirteen-block, you could use that too! We call thirteen *prime*. The primes are the Numberfolk that can only be built with ones or themselves. With the primes we can create every counting Numberfolk."

"I get it. So, none of the even numbers can be prime, because they all can be made using only two-blocks."

"Almost! We saw that four is not prime. We call the not-primes composites, but what about two?"

"Oh. Two can only be made with one-blocks or a two-block. Does that mean two is prime, too?" Marco smiled. He enjoyed saying 'two too'°.

"Exactly! Two is the only even number that is prime. All other primes are odd." Mr. Pikake held up two fingers. He was starting to remind Marco of Count von Count from *Sesame Street*. "You have learned two new terms: prime and composite." He slowly raised a third digit, "Multiple. Multiples are all the stacks that can be constructed with certain blocks. You have already marvelously mentioned the even numbers. They are all multiples of two as they can be built using two-blocks. What are the multiples of three?" The ball was in Marco's court.

Luckily, Marco was skilled at counting and quickly retorted, "3, 6, 9, 12, 15, 18, 21…."

"Perfection!" Mr. Pikake interrupted, which came as a relief. Marco wasn't sure how long he'd have to continue counting to appease the man. "Composites have multiples as well. For the multiples of four are 4, 8, 12, 16, and on and on. But," this pop was louder than the previous, "we aren't so interested in these. They are all also made of twos. We want to rip apart any Numberfolk into their…" He stopped midsentence to outstretch his arm, nearly bopping Marco in the head as his fourth finger sprung to life, "factors. Our fourth and final term."

° Try it, you'll smile too.

Breathe Marco, breathe, he thought as his rollercoaster roared back to life. The fourth term was factors, he got that, but he had already misplaced one through three. What were they again?

Seeing the change in his student's expression, Mr. Pikake leaned in and whispered, "Don't fret. Remember we are candidly counting. Never underestimate the power of the count." He flashed a crazy smile.

I can count. Marco reassured himself before asking, "What are the four again?"

With the excitement of a child, the professor recounted, "One. Prime. Two. Composite. Three. Multiple. Four. Factor." He chuckled then added, "I prefer Primary Prime, Consequently Composite, a Medley of Multiples, and Finally Factors."

Marco took a moment to catch up. Bowing his head and pushing his face into his palms, he rubbed his eyes. *Primes are numbers that can't be split evenly into groups, composites are all the rest, and multiples are the numbers you get when you count by something. I can do this.* Feeling ready for factors, he sat up giving Mr. Pikake a quick smile and a nod.

"Ready?" the professor asked to make sure. Marco nodded again. "Wonderful! You will need the power to rip apart the Numberfolk at the seams. We are almost there," he winked. "Factors are all the clusters, the blocks, that can make up a Numberfolk. Since six is made up of two three-blocks, it has factors of two and three. You try now. What are the factors of twelve?"

The power to rip something apart was just what Marco needed. Making quick work of his imaginary LEGO set, Marco stacked blocks of size three until he had twelve. It took four three-blocks so, "Three and four?"

"Amazing!" the tutor shrieked. The 'A' was quite loud, but he caught himself and pulled the 'mazing' down a few decibels. "Now, this tells us that twelve is a multiple of three and four *and* that three and four are factors of twelve. We can count to twelve by threes or fours – multiples, and we can rip twelve apart into three fours or four threes – factors. But twelve is not only a multiple of three and four. Anything you can count by to get to twelve also claims twelve as its multiple."

He pushed back his chair and began circling the table chanting in a singsong voice, "We can use ones or twos. We can use threes or fours. We can use sixes, or we can use a twelve! This means twelve is a multiple of 1, 2, 3, 4, 6, and 12 *and* 1, 2, 3, 4, 6, and 12 are factors of twelve." Making his way back to his seat he threw himself down. Placing both palms on the tabletop he added, "They are all factors, but our goal, boy…our goal is to break them apart into their most basic forms. To see inside their head. To know what makes them tick, exactly who they are. And then, then we will dominate them!"

A bit confused, 1, 2, 3, 4, 6, and 12 were dancing – no swirling around in Marco's head. Were they multiples? Or were they factors? Maybe both? "Are factors and multiples the same?" The words left his mouth without permission.

"A common point of contention!" The professor's voice started strong and became softer as he continued. "Likely because they are both the same and different." *That's super helpful,* Marco thought. Mr. Pikake cleared his throat. "The multiples of a Numberfolk are never smaller than it. The multiples of 12 are 12, 24, 36, etcetera, etcetera. The factors, however, are never bigger! The factors of 12 are 1, 2, 3, 4, 6, and 12. Where they are the same is any Numberfolk that has twelve as its multiple, must then be a factor of twelve! So, to break down twelve, you may consider: what has twelve as its multiple?"

Marco was shot to an interrogation room. It was dark and empty except for a cold metal table and single light eerily swinging overhead. Playing bad cop Marco yelled at the twelve, 'Tell me everything!', slamming his fist down.

'Okay, okay,' the twelve whined. 'I work with one and two, but I swear that's all!'

Marco slid a file across the table. 'We know you're lying! We have seen you with three and four! And there you are with six!' The daydream was enough to get him excited about the idea. He was ready to tear apart some numbers.

Mr. Pikake pulled a thick sheet of paper from his briefcase and wrote the number twelve in large strokes across the top of the page. "Let us get to work." With fiery eyes, he drew two lines extending from the twelve. "Rip it apart, boy! It doesn't matter how. Pick a way to split the twelve into clusters. What size blocks will you use?" Deciding to stick with the four three-blocks he used earlier, Marco sketched a three and four beneath the lines.

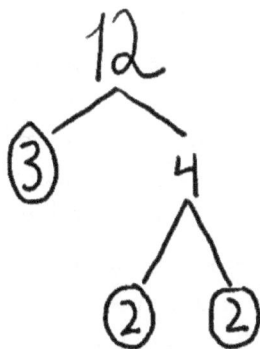

"Splendid! But," a pop, "we know there are no four-blocks, so you must split that, too!" Marco obeyed. He drew two lines from the four and under each wrote a two. "Perfection! Once you have the Numberfolk figured out, split entirely, circle the primes, the indivisible blocks. Then, you will have broken it down, you will know what it is made of, everything it is made of, and then you will have the ability to control it!" Marco circled the three and each two. As he did it, he felt strong, he loved the idea of dominating the numbers.

"You've done it, my boy! Now, you know the twelve is nothing more than a three, a two, and another two. You have shredded it into its simplest pieces. You can see inside its mind, what drives it, and you can make it do whatever you wish.* Take two blocks each

---

* They call this Prime Factorization, and it simply means to break a number down into its most basic building blocks. If you multiply the prime factors of a number, the result is precisely the number. Notice $2 \times 2 \times 3 = 12$.

of size two and stack them, which makes a single block that is size four. Now, use three of those and you will have twelve."

"But..." the word came out of Marco's mouth before he was ready for it. He had no choice except to continue. "Doesn't it matter that I picked three and four? Wouldn't it be different if I tore it into six and two first?"

"You tell me." He slid the paper gently in the student's direction. "Split it in half instead now." Marco wasn't sure he could do it alone. The doubt was intense, but the power was intoxicating. He wanted more. Drawing his own diagram, he split the twelve into two and six and circled the prime number two. Then he broke the six into a two and a three, both primes, two more circles. "It is the same!"

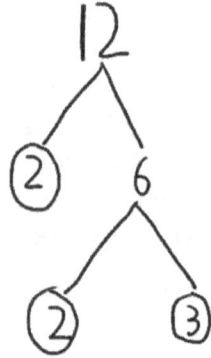

"Dazzling deduction! You see the twelve is who it is, it does not matter the journey you take or the questions you ask it. Ultimately, when you tear it down, you will see into his soul and then you will know him. Now, maybe you take two blocks of size three to make a six-block. And two of those again make a twelve."

He imagined ripping Peter apart like the numbers – looking deep into his soul to find nothing more than a whiny baby. As Marco looked over the page, he saw twos and threes, in one of the images there was a four and in another a six. All twelve's factors were somewhere except ones, they were nowhere to be found. "I thought one was a factor. Where is it?" he asked his tutor.

"Fracturing." Mr. Pikake said the word in a magical tone, like the *abracadabra* that came at the end of a trick. "Every Numberfolk can be fractured. If we rip twelve into single blocks, we have, well, twelve blocks. We know nothing more about the creature. You see, they say that one is neither prime nor composite."

$$2 \times 2 \times 2 \times 2 \times 3$$

"Isn't it prime?" Marco protested, "I mean, a prime can only be built from ones and itself, and one can be built with one one or one one." The doublespeak made Marco dizzy.

"You have made the argument for me!" The professor smirked. If Marco had made an argument, he wasn't aware. "One one and one *one*," Mr. Pikake attempted to differentiate the two by accenting his final word, "are the same." He slid a single M&M across to Marco. "Can you make two different groups with this?"

Marco didn't understand, he couldn't do much of anything with a single M&M other than eat it. Grabbing another chocolate the tutor continued, "Look here. I can split these two candies in two ways. I can make one group of two or I can make two groups, each with only one. Can you do the same with only one M&M?"

Now Marco understood. One couldn't be made up of other clusters, there was only a single way to make one – no matter how you looked at it. "Oh, I see. Primes can be split in two ways. One way is into groups of singles."

"Fracturing," Mr. Pikake chirped.

"Yeah, and the other is into one group with all of them. But one can't make two groups, just the one that is really both of those."

"Precisely! One is made up only of itself – not itself *and* ones, so we cannot call it prime. We certainly cannot call it composite. And if we rip everything into singletons – fracture them – we know nothing. Do you see?"

Marco was back in the police station. As he tried to piece together a profile of the suspect, a fellow officer ran in shouting about new information. Marco snatched the paper from his hand and read the bulletin: We have confirmed the suspect is a number. *That doesn't help at all*, he thought smiling. "All the counting numbers can be fractured, split into ones, so ripping them this way doesn't make much sense, it doesn't tell us anything about them."

A large devilish grin grew on the professor's face as he took the page and began scribbling. "Show me once more your skill." He tapped at the page revealing the number forty-two. "Who is this, Marco? What are they made of? And don't say ones. We're all made of ones," he chuckled.

He was in a courtroom now. Forty-two sat next to the judge. Marco whisked across the room shouting, "Do you deny you gave the job to two twenty-ones?!" Forty-two hung his head unable to look Marco in the eye as he nodded. Turning to the judge Marco demanded, "I call my next witness, Twenty-one, to the stand." Twenty-one made their way across the courtroom. Marco noticed something distinct about it. It had three legs, well, it had two legs and a cane. The instant the number sat down Marco turned, "The crime scene was unique, it showed markings of a three-legged man. Was this not you?!"

The number was defiant, "I have three legs, sure. But that isn't enough. Six has got three legs too. Why d'nt you question him?"

Turning to the judge Marco said, "If it pleases the court, I'd like to ask the witness to show us their leg."

Approving his motion, Twenty-one made their way to the center of the room and lifted their pant leg. A gasp rang out from the jury as each of the number's legs were like an insect, segmented into seven pieces. Twenty-one burst into tears.

With a smug look Marco concluded, "As you can see ladies and gentlemen of the jury. It was none other than two, three, and seven who committed this heinous crime! I rest my case."

He completed his drawing and pushed it to Mr. Pikake. His rollercoaster was flying now, so fast it nearly jumped off the tracks. But Marco wasn't scared, he was crazy with excitement. His hands flew up as he enjoyed the ride°.

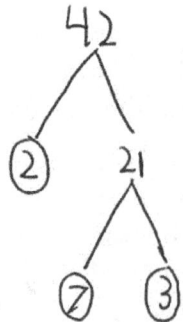

° Try it out! Can you find the prime factorization of 50? How about 121? 17?

2 × 5 × 5

"I knew you were special," the professor said calmly. "You have completed your first training – knowing the enemy is key to any battle. When you know what you are working with, when you can control the Numberfolk, well, Marco, *then* you can control the world."

Normally, Marco wasn't the one to raise his hand in math class. He wasn't the type to ask questions. He considered himself a Lost Boy, often abandoned and confused as the class moved on. But something about the professor made him feel safe. Not only that, Marco desperately wanted to feel more of that power. He needed to. He had directly disobeyed Peter, and this had to be worth the wrath that was sure to come his way. What was really bothering him were all the letters in his math class. "Can you rip apart the letters, too?" he asked.

"*Ah*. The letters." Mr. Pikake stroked his chin. "I intended to investigate these next time, but I suppose a swift scan is alright." With a wave of his hands the M&M's were back between the two, scattered everywhere. "While we can manipulate Numberfolk, uncover what makes them tick, they are excellent hiders. They are always trying to cloak their true nature, trying to fool us." Spit flew from Mr. Pikake's bottom lip as his precise pronunciation of each word made 'fool' come hurling at his student. With another wave of his hand, Mr. Pikake had rearranged the candies. This time into four large X shapes. He continued speaking faster, yet still with a cutting clarity to each word. "Do you see the letters, Marco? Do you see them?"

Marco nodded silently. He was terrified and intrigued of both the letters and the tutor.

"What letters do you see?" Mr. Pikake's head rotated like an owl searching the room. Locating what he was hunting for, he sprung up and in four long swift strides, grabbed the chalkboard that sat against the back wall blocking the windows and pulled it to their table.

"X," Marco choked out.

"Yes!" the tutor shrieked, so enthralled with their conversation he didn't even bother looking to the librarian. The chalk as his sword, he made two elongated slashes on the board.

"And how many do you see?"

"Four." Mr. Pikake battled with the board again to reveal four Xs.

"Now remember," he was still talking at an increased speed, "they can hide, they can try to deceive us, they can twist and bend themselves into many different forms. They are shapeshifters but YOU, you are the commander, the detective, the hunter. The number has changed its form, but we know who it is. Who is it, Marco?!"

Marco stood. He had seen this phenomenon before at pep rallies and in movies when lieutenants gave moving speeches to rally the troops. Mr. Pikake's excitement was viral, and it had infected Marco. "It's 68!" he cried.

"PRECISELY!" With that, Mr. Pikake finished his scribbling and, in a flash, had returned to the table. Marco looked up at the chalkboard to see:

$$X + X + X + X = 68.$$

Feeling sick from the mix of excitement and fear, Marco shifted his eyes back to the table, to gain his bearings. The four Xs on the board equaled 68, just as the four large Xs on the table were built from 68 candies. For the first time, he began to understand how letters and numbers could work together. The rollercoaster was moving at full speed now. He had gone through fear, apprehension, and excitement in a matter of minutes. Now, something new lay ahead. It began to swell inside him, quickly taking over his entire body. It took a moment for Marco to realize what it was – it was hope.

"When we first met, you shared that the letters were thwarting you. Don't let them, Marco. Letters are nothing more than numbermasks. One more way for Numberfolk to hide, to try to deceive us. Who is hiding behind the X? Who is trying to trick us? Don't let them. We cannot be tricked!"

Feeling invincible, Marco examined what lay in front of him. Each X was exactly the same. They all had four arms, built of four candies each. One additional chocolate sat right at the center of the X connecting everything together. Quickly completing the addition, Marco reasoned that four arms, each with four candies made 16, and including the centerpiece, the makeup of each X was 17 M&M's. He started to feel a tad maniacal himself, but he didn't care. Something finally made sense. He allowed the feeling to take over as he shouted, "17!"

Head Librarian Sabrina had enough and was not so kind to the boy. She stood, her arm snapped out like an arrow being fired from a bow. She was pointing to the door.

With a chuckle Mr. Pikake whispered, "I guess we should get out of here!" The two grabbed their things and made their way quietly to the exit. Once in the hall, Marco was the first to speak. He bounced up and down childlike asking, "Was I right? Was I right? Was the answer 17?"

"You are astoundingly accurate!" Mr. Pikake beamed. "Your teacher would say, 'together four Xs make 68, what is X?' But we know this to mean only that a Numberfolk is hiding behind a numbermask. We must use our powers to rip apart and put together 68, to hunt down who is hiding, who is trying to deceive us. You are a master counter, Marco. You can count by 2's, 3's, 4's, 5's, 10's, 17's, anythings! Now you, too, know how to classify the enemy into prime or composite, to find multiples, to find factors. Most importantly, you have stared them down into their souls and determined everything they are made of. Your counting is a great power to uncover who's hiding. And now? Well, now you are beginning to understand how to use it."

As they left the school, Marco was relieved to see his mother's white sedan parked out front. His sister's unmistakable curls were bobbing up and down in the back seat. He didn't think it was possible for Maggie to sit still.

Marco thought he saw a look of disappointment cross Mr. Pikake's face. If it had been there, by the time Marco blinked it was gone and had been replaced with a kind smile. "Excellent work today, boy. I look forward to our meeting tomorrow."

Marco ducked into the car just in time to hear Maggie's rant of the day. "So, she had these mosquitos in jars and passed them out to everyone. We were supposed to be studying the parts of their body to find the head, thorax, and abdomen. But a bunch of kids were trying to take the lids off to release them. Can you believe that?! You should write a letter to the school or something. I mean, doesn't she read *National Geographic*? Do you know how many people die every year from mosquitos? It's practically child abuse to have these in a classroom. Not to mention all the inhumane treatment of the mosquito issues – shoving them in a jar and all. I wouldn't be surprised if tomorrow there are black widows or poison dart frogs on everyone's desks." She then turned her attention to her brother, "Was that your tutor? You look like a toddler next to him. That can't be good for your confidence."

Maggie tried to hit Marco where it hurt. She knew he had always been self-conscious about his height. Both Liam and Oliver stood almost a foot taller than Marco, and he was one of the shortest in his grade. One year Marco had adopted a spikey hairstyle that made him appear three inches taller to compensate. However, in that moment, not even his sister's jabs could get to him. Marco felt amazing. "I'm still taller than you, aren't I?" he stuck his tongue out at her and made a silly face. Nothing was going to bring him down.

"How was school today?" his mother asked.

"Good. Really good," he softly replied allowing the smile to wash over his face. He didn't even attempt to hide it, and he gladly allowed it to stay there the whole ride home.

KNOWING THE ENEMY UNIT 1

I debated whether or not this counted as a "blank page". Logically, *if* the page is blank, *then* there is nothing on it. Hence, by the contrapositive, *if* there is something on the page, *then* it is not blank.

That means mathematically this was not a blank page, but I wasn't sure everyone would compute the logic to convince themselves of that.

Anyhow, it's a good time to pop-in because you may be thinking *"What is the Kryptografima?"* The Kryptografima is an online school created by the society Mr. Pikake works for, but you don't really hear about that until Chapter 7. Without giving anything away, I'll tell you this: The Kryptografima contains games and quests that allow you to grow your skills just like Marco. For instance, you can hop on now and practice ripping apart Numberfolk.

One more thing before you continue (the next page by the way is one of my most favorite). If you haven't noticed yet, Numberfolk have taken over this entire book including the bottom of each page. Have you detected their pattern?

$$1^2 + 2^2$$

$$25\% \text{ of } 20$$

$$\frac{325}{13}$$

$$5 + 5 \times 5 \div 5 - 5$$

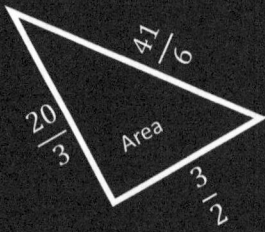

$$f_5$$

$$\sqrt{25}$$

$$\frac{20}{3} \quad \text{Area} \quad 4\frac{1}{6} \quad \frac{3}{2}$$

5

LETTERS

"NATURE IS WRITTEN IN
MATHEMATICAL
LANGUAGE."

-GALILEO GALILEI

On Wednesday, Marco woke up feeling like he had a new lease on life. He was a new person. While he wouldn't go as far as to call himself a 'math kid', he didn't consider himself *not* one either. He was walking the line and the high-wire act was invigorating.

Although his tutor bordered on insane, he was looking forward to their session. There was something fun about insanity. It was freeing, interesting, different. Mr. Pikake talked about numbers in a way that was entirely foreign to Marco. Yet, seeing the X of M&M's, understanding that numbers are just hiding and it is our job to find them, ripping them apart to see what they were made of…well, that was a language Marco not only understood, but enjoyed. The professor was turning math into a sport, into a hunt, into a place where Marco was the hero.

As he made his way downstairs for breakfast, Marco found himself standing a bit taller than normal. He felt strong, he felt capable. He was relishing the feeling when suddenly a powerful force from behind sent him flying forward. Barely catching himself on the landing, another jerk came and he was tumbling to the right. The office door slammed as Marco tilted his head to see Peter.

"I thought I made myself clear. Put a stop to this tutoring." Peter's deep brown eyes blazed with anger.

Not sure if it was his new command over numbers or maybe just spending time with Mr. Pikake, someone who built him up rather than tore him down. Whatever it was made Marco feel differently. He wasn't scared of his stepfather. Slowly pulling himself off the floor, still looking down Marco softly – but firmly – replied, "No Peter."

"What did you say?" Peter's fist clenched.

"I said no." Marco was staring directly at Peter now. "I like my tutoring. I like my tutor. Mom wants me to do it, so I am going to."

"Well, you think you got things all figured out, huh? You think you can just disobey me and nothing will happen? You just wait." He threw his shoulders forward like he was going to attack but his feet remained firmly planted on the ground. Marco recoiled. The grin that grew on Peter's face told him that was exactly what his stepfather wanted. "I am an exterminator. That makes me an expert at catching things nobody wants around. That makes me patient. You do what I say or when you least expect it, you'll find that I *will* get my way."

His nasty smirk transformed into a friendly smile as he turned and left the room. *The many masks of Peter,* Marco thought. He had never been defiant before, and he liked how it felt. His stepfather didn't get to decide his life. Yet, as much as Marco tried to convince himself he wasn't scared, a horrible feeling bubbled deep in his gut. He repeated to himself, "Numbers are like people," the entire way to school as he imagined hunting the Peter-mask and dominating it.

Despite the enjoyment his tutoring sessions brought, he had not yet fallen in love with the educational jail known as Mrs. Sanders' class. The beast of the day was word problems. In an unusual change of events, as his classmates struggled to translate sentences to numbers and letters, then to numbers, and finally back to words again – for the first time, Marco found himself at the head of the class.

"A new robot toy was released, and you are excited to buy it." Mrs. Sanders always tried to make word problems have some interest to her students. While it was a nice thought, they all boiled down to an equation of some sort with all the context removed anyway. "It is listed for $50, but there will be a sale offering 20% off. How much will you save if you buy the game on sale?"

Oliver's hand shot up. Marco's best friend, like his sister, possessed a genetic supernatural power that made them number savants. While everyone else struggled to move an inch, they could come in and easily jump a mile. The fact that Oliver was also the class clown and always in trouble helped balance the nugget of jealousy that ached in the pit of Marco's stomach.

"Yes, Oliver?" Mrs. Sanders pointed to her student. "You will save 20%," Oliver answered with a smirk. Quiet giggles erupted from all corners of the classroom.

After a sharp look, Mrs. Sanders painted her smile back on and responded, "You are correct. But how much *money* would that be?" In an instant, the class became the robots from the problem and following their programming, they all put their heads down and began scribbling on the paper in front of them. Unlike his cyborg peers, Marco put his head down, but wrote nothing on his paper. He had a different idea. Remembering what he had talked about with Mr. Pikake – that numbers are ours to control – he let his imagination take over.

Knowing that 'percent' meant 'out of 100', Marco saw the number 100. A vacuum cleaner was sucking out 20s from the number. *SLURP!* Now, it was an 80 and a 20. *SLURP!* Now two 20s and a 60. Marco kept sucking at the number until it lay on the ground as five disoriented 20s. He turned his attention to the $50 from the question. He changed the settings to rip the 50 into five

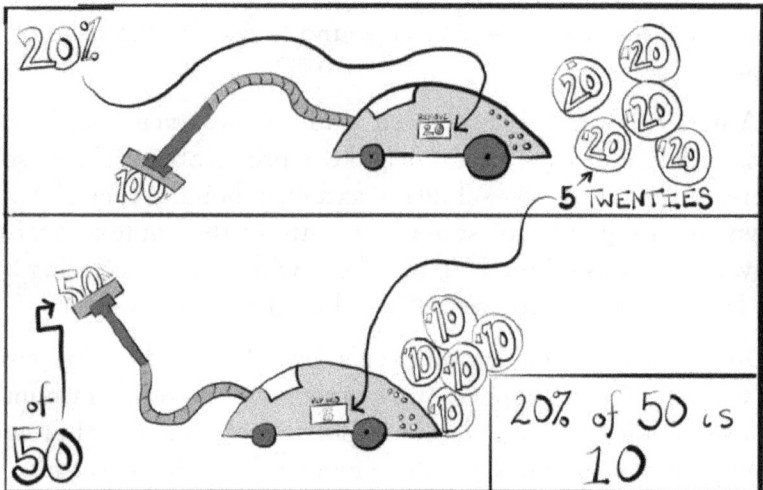

pieces as well in one mouthful. *SLURP!* Five 10s shot out the back.°

He slowly raised his hand. Oliver shot Marco a look that read 'what are you doing?'

"Marco," Mrs. Sanders called.

"10?" Marco answered reluctantly.

Mrs. Sanders looked at the clock which read 9:58, "No, it's not quite 10 yet," she responded.

"No. The question. You'd save ten dollars."

Her face frozen in a state of shock, Mrs. Sanders looked down at her paper and back up at Marco. "Oh, *um*, yes. That is correct, Marco. Nice job."

Riding his math high for the rest of the day, Marco was silently thrilled when Oliver took it upon himself to recount the event for Liam at lunch.

"You should have seen it, man!" Oliver exclaimed. "The teacher couldn't even solve it as fast as Marco." He turned to add, "Your tutoring must be going well."

Marco pushed and pushed but couldn't hide his smile. His green eyes were bright with excitement as he began to tell his friends about Mr. Pikake. "He's a little crazy, I'm not going to lie. It makes things fun. It's like a video game, but rather than zombies, we are

---

° What did Marco do here? The question Mrs. Sanders really asked was 'What is 20% of $50?' Since five 20's make 100, he needed to know 5 of what would make 50. He split 50 into five pieces to discover each piece is a ten, meaning 20% of $50 is $10. You can do this with any question where the percentage is a factor of 100. My dog ate 12 pounds of food last month and he ate 25% more this month, how much did he eat? Since it takes four 25's to make 100, how many fours make 12? Well, three of course. So, 25% of 12 is 3. Fido ate his normal 12 pounds plus the 3 more meaning he ate a total of 15 pounds of food this month – what a pig!

hunting numbers." Pretending to be a character in the boy's favorite game, Marco mimed out a zombie hunt.

"Are they evil numbers trying to eat your brain?" Liam laughed. Although Marco hadn't thought about it before, the way Mr. Pikake talked about numbers gave him the sense there was something dangerous about them. Realizing how bizarre that was, he shook the thought from his head.

"Too bad you can't figure it out yourself like the rest of us," Oliver snapped. Marco was confused, he shot Oliver a puzzled look. "You think you're so big and mighty because you solved one problem. Get over yourself." He pushed himself back from the table, grabbed his things, and stormed away.

"What was that about? I thought he was happy for me?" Marco turned to Liam.

"Don't worry about it. Math was his thing. He'll get over it."

When the final school bell rang, Marco practically jogged to the library. His tutoring was becoming an ice cream eating contest. The more he ate, the more it hurt. His stomach, nauseated from Peter's threats. His brain freeze pounded as he tried to figure out how to mend things with Oliver. Despite the pain, the sweet treat tasted so good he couldn't help going back for more.

The professor was waiting at what now felt like *their* table. Marco gave his tutor an excited wave and started to shout out 'hello.' The feeling of Head Librarian Sabrina's eyes on him made Marco pause. He wisely decided to quietly walk over instead.

"Good day, Marco!" Mr. Pikake beamed.

"What are we doing today?" Marco asked. "If we use M&M's again, I think we should at least get to eat them at the end."

A light switch in the professor's head flipped, turning Mr. Pikake's expression serious. "Before we get started, it is my sworn duty to tell you more about Numberfolk. I don't tell many people this, Marco." He scanned the room to ensure no one could overhear them. "But, I trust you. I know you have greatness in you."

Marco nodded with respect, trying to show Mr. Pikake that he was right to trust in him.

His tutor continued solemnly, "Numberfolk, well...they govern the world. They do so invisibly, secretly, and so well most people don't even realize it. Numberfolk dictate how many petals a flower will have, how hard gravity pushes down on us, how a ball will fly through the air. They are behind everything we see and everything we don't see." He took a deep breath before going on. "The good news is Numberfolk are predictable and *controllable*. This is what gives us the upper hand. As long as we can recognize the patterns and understand how to utilize our powers, we can be victorious."

Not knowing the reaction the professor was expecting, Marco only blinked. He felt like he was being inducted into one of Liam's secret societies – one of numbers. Despite his friend's warnings, at this point, he didn't care. As long as he kept having wins in his math class like earlier today, it was worth it. He'd figure out Peter and Oliver later – the power was too much to turn down. He forced a smile and a nod.

The light switch flipped again, and Mr. Pikake was smiling as joyfully as ever. "Today, let us talk more about those letters that are giving you so much trouble!" Marco sat up and listened carefully. He hated the letters, and this was his best chance to finally begin conquering them.

"Yesterday, we saw how letters are just numbermasks, allowing Numberfolk to try to trick us." His t's packed a powerful punch. The professor reached into his briefcase and removed a piece of thick stationary which he sat on the table between them. "First, don't let the letter fool you. They may be hiding behind an a, or a b, or a c, an x, or a y, or a V. Really, they are all just Numberfolk."

With long smooth strokes he wrote on the stationary:

$$a + 7.$$

"We begin with one part, then will add two. This here," he pointed to the paper, "this is an expression. It is how the Numberfolk express themselves, how they appear to us. Expressions can be simple or complex costumes meant to deceive." He looked down at his student before going to work attacking the parchment.

$$3m - \left(4 + \frac{m}{2}\right) + (-3 - 6m) + \frac{1}{2}(m + 2).$$

Marco's eyes were wild with interest. Just behind the glitter lay the shadow of doubt and the fire of fear. Seeing this, Mr. Pikake placed his hand in front of his lips and whispered, "Don't worry, we will start simple. An expression only means there is no equal sign. We are looking at just one side of the Great Scale." Appreciating the reassurance and clarification, Marco smiled. His tutor returned a jovial wink.

"The ability to equate is where the real power lies. What if I said to you, *'pias to avgo kai kourefto'*, could you decode what I mean?"

This is how Alice must have felt. Marco thought. *The professor is the Mad Hatter, and everything is nonsensical!* Normally, Marco's confusion would mean giving up. However, his tutor had lit a spark inside that was beginning to rage – filling him with determination. "I, *uh*, don't speak *that*?"

"I don't either! Hopefully it isn't anything offensive," Mr. Pikake laughed and Marco joined in. "What I am trying to say is that an equation is like a translator. Without the equal sign, we don't have anything to hold on to, nothing to help us decipher its meaning. Both sides of the Scale are required to hunt. But," the professor popped, "Numberfolk are predictable, which means we can play with an expression to reveal its patterns. And that…that can be *very* powerful."

As Marco began to catch on, stress started melting off his shoulders. He was at a crowded bus stop and person after person kept bumping into him. When he looked closer, he noticed they weren't people at all! They were numbers dressed in suits and dresses, carrying bags and umbrellas. He tried hard to listen in on their conversation, but it was pure gibberish. He pulled a device out of his pocket and looked at the screen, on which were two horizontal lines: =. He placed his finger on the symbol and suddenly the voices changed. "How are you doing today, Six?" "Swell Eleven, how's the kids?" He giggled to himself.

"Now, let us push to pinpoint the pattern!" His slender finger motioned to the $a + 7$ on the page. "If 5 is hiding behind a, what would *we* see?" The way Mr. Pikake said 'we' made Marco feel like the two of them were special, they had a secret power that others didn't know about. He looked down at the paper, a 5 peaked out from behind the a and giggled before disappearing again.

"5 + 7?" Marco guessed.

"Yes, yes, but that is a duel. We'll learn more about that later. *Who* is $5 + 7$?"

"Oh 12, that's an easy one."

"Precisely, lad! Now, the power in this, which we call *substitution*, is that we can begin to understand the rules Numberfolk follow." And there it was. Rules. Marco's internal fire was instantly extinguished. He hated rules. Noticing the change in his student, Mr. Pikake added, "It's good the Numberfolk follow rules. It allows us to better know them. Especially if they are rational, just wait 'til we get to the irrationals. Anyhow, if they are following the rule $a + 7$, when we see a 12 we know exactly who is hiding behind a, you see! Know your enemy!" Like a guard dog, he barked his last words. "Have you ever studied World War II in school?"

Appreciating the change in topic, Marco replied, "Sure we have."

"Do you know that in the war, the Germans used a device called the enigma machine?"

Marco shook his head indicating to Mr. Pikake to continue, "When the enemy can decode your patterns, they know the rules you play by, they understand what will come next and this gives them an unbelievable advantage. The enigma machine was a device that coded communications between Germany and their allies. If no one could understand the messages, then no one knew their secrets. Once Poland cracked their code, victory came more easily." He tapped the sheet of paper, "You are looking at a number code, Marco. Once you know how to decipher the code, you too can win the war."

Marco looked back at the paper with a new perspective. It wasn't a math problem – it was a puzzle to solve. He could latch onto this idea...he enjoyed a good mystery.

"Let's pick another Numberfolk to hide. Say this time it's 15, what would we see?"

The 5 peaked out again from behind the letter. This time, it had a giant 1 attached at the hip. The 1 looked angry, pulling himself and the 5 back down behind a. "If 15 is hiding," he continued in his head, *then we would see* $15 + 7$ *which is*, "22!" he exclaimed.

"Not so scary, ay?" Mr. Pikake grinned. A burst of pride swam from Marco's stomach up to his throat. "By imagining what numbers might be hiding, we can now decipher the pattern the Numberfolk are following. For example, you are on the front lines, and you see the Numberfolk sending in larger and larger soldiers to the $a + 7$ machine, what would you expect to see come out of the machine?"

Mr. Pikake had done the hard work for him. Marco imagined a big factory. Large chimneys were spitting out plumes of smoke and the expression $a + 7$ was spray painted on the side of the building. Numbers lined up on one side of the factory, there was $37, 38, 39$! One-by-one they entered the building, it churned and popped until $44, 45,$ then 46 came out the other side. Not only were the numbers larger, they were beefier, too. The $a + 7$ must be a super soldier creation factory.

"When the numbers going in are bigger, the numbers coming out keep getting bigger and bigger, too!" Marco didn't feel like he was catching on, he *knew* he was starting to get it.

"Perfection Marco! Now imagine they are following the rule," he finished his sentence by scrawling the next expression across the stationary:

$$20 - x.$$

"If 4 is hiding behind x, what will we see?"

"We would see 16! $20 - 4$ is 16."

"And if 10 is hiding? Then what?"

"We would see 10!" Marco thought that one was a little funny. *It isn't a very good hiding spot if we can see you, Ten.*

"You've got it now! And what pattern do we have here? If the Numberfolk keep sending in larger and larger values to hide, we will see..." Mr. Pikake paused to allow Marco to finish his sentence.

"We see smaller and smaller values. The more we take from 20, the less is left."

Mr. Pikake jumped from his seat and motioned for Marco to follow. The two pranced between the library stacks. Mr. Pikake waved his arms as he threw expressions and values to his student. For Marco, it was a baseball game. The tutor would pitch a question and Marco would hit it out of the park.

"What if the pattern is $c + 6$? If 8 is hiding, we shall see whom?"

"14!"

"And if 10 is hiding?"

"16!"

"What if 22 is hiding, my boy?

"We'd see 28!"

"Yes, yes! And if the pattern is $x - 6$? Tell me what happens, lad!"

"If 10 is hiding, we see $10 - 6$, so 4. If 14 is hiding, we see $14 - 6$ or 8. If 92 is hiding we see $92 - 6$, so 86!"

They arrived back at their table both laughing. They made sure to keep it quiet to avoid Head Librarian Sabrina's wrath.

"So, it doesn't matter what number you guess? You can just say any number is hiding?" Marco understood the idea of substitution, just not its use. It seemed like they were just making things up, which was fun and all, but he didn't remember ever being asked to make up a number on a test.

"Excellent inquiry." Mr. Pikake wrapped his thumb and his pointer finger around his chin. "We are working with expressions meaning, no equal sign. If we just have $x - 6$ and don't know what it is equal to, anyone could be hiding behind x. On math tests," he said the word *tests* with a clear disdain, "teachers may ask you questions like we practiced today in coded terms. They will say 'Evaluate the expression $x - 6$ when $x = 7$' or some other nonsense. You calculate $7 - 6$ and draw a nice circle around the 1." Marco couldn't help but laugh at the way the professor talked about teachers and tests. His tone implied feelings of contempt that Marco could relate to.

"The expression has two main powers. When we test the expression to see what happens when different Numberfolk are hiding, we can determine the pattern they are following. Some say mathematics is precisely the study of pattern! It allows us to predict what will happen next and to better understand the forces that are controlling everything in our reality. We will only dabble. Higher masters who study calculus, behaviorists, well they can simply look at an expression and dictate how it will act, how it will behave forever. We shall stay small. Tell me, what is the pattern of $x - 6$?"

"As bigger values hide behind x, the $x - 6$ is getting bigger and bigger. Because $6 - 6 = 0$ and $7 - 6 = 1$ and $8 - 6 = 2$ and they just keep growing!"

"Perfection! Patterns can be deceiving, too. You have heard the tale of Hansel and Gretel, yes?"

Marco nodded.

"Well, when the children first arrived at the witch's home, she made them a deal. She presented each child with a candy bar and explained that every day, she would give them more candy. The amount would be exactly half of what she gave them the day before. If they could resist eating the candy, then when they returned to her two full bars, she would set them and their father free. Does this sound like a good deal?"

Never having heard this part of the fairy tale, Marco thought on it a moment. *If every day she is giving them more candy, then their candy will grow and grow and grow.* "I think so. They keep getting more so they will eventually have two bars, even more than two bars!"

"*Ah!* The children thought the same thing! They agreed to her proposal, and she allowed them to take the full bar that day. The next day, she gave them one-half a bar. Hansel and Gretel thought this was marvelous. They were already halfway to freedom! They had their first bar, and half of their second bar. The following day, she gave them one-fourth of a bar. Feeling like their release was near, the children rejoiced. Like you, they saw the pattern. They were getting more and more each day. But alas, this continued for many weeks. Next, she gave them one-eighth, then one-sixteenth, then one-thirty-second. Each day half as much as the day before, just as she promised. The children saw their bars growing and growing, but by smaller and smaller amounts.°"

The professor took a long pause. Too long. "Well…what happened?!" Marco prodded.

"Sadly, the children never won their freedom." Mr. Pikake said with a sigh, "You see, although the pattern was telling us they were gaining more and more candy – which they were – they would never amass two full bars."

"Why not?" Marco was not convinced, if they kept getting more why wouldn't they at some point have what they needed?

"*Ah*, you see, patterns, while powerful, can be quite deceiving. By giving the children half the previous amount each day, it meant they would never get the second full bar, at least not in a lifetime! Why? Well, because they were adding such a small amount each day, it would take forever. It was both a possible and an impossible task. What we call a paradox."

Marco imagined the little children, only able to receive crumbs, less than crumbs, so small they appeared invisible. Tears dripped from their eyes.

Mr. Pikake continued, "So you see, understanding the patterns, truly understanding them, is a very formidable strength! The second use of the expression is similar to the enigma machine. It allows us to translate. You know these as…" his voice trailed off.

"Well, as what?" Marco pressed.

Mr. Pikake looked around nervously, "The f-word." He paused briefly before continuing in a low, deep voice, "I don't generally say the word. They are nasty things, *formulas*." The f-word flew out of his mouth like a butterfly who just emerged from its chrysalis. It floated in the air a moment and lingered like a bad

---

° The children were given $1 + \frac{1}{2} + \frac{1}{4} + \frac{1}{8} + \frac{1}{16} + \frac{1}{32} + \cdots$. This will eventually equal two. The problem is, it would take longer than forever to get there.

taste. The professor clenched his teeth as if shielding his mouth from the dirtiness of the word.

Marco *hated* formulas. They were absolutely the very worst thing in math. Formulas were rules you had to memorize, but rather than plain English like 'do the parentheses first' they were cryptic messages no one understood[*]. "I hate formulas," Marco muttered.

"As you should," Mr. Pikake boomed. Marco was flabbergasted by his tutor's response. He had never heard of a math teacher who didn't love a formula. "Many great masters before us derived formulas by doing just what we are. Looking at the Numberfolk and finding ways to describe their patterns. Don't worry, I will never require you to memorize an f-word. What a waste of a precious brain! We shall instead understand what the great minds understood and therefore understand ourselves. Then, we never have to memorize, we will just know. We shall stand on the shoulders of giants and see farther than we could on our own two feet!"

The idea of never memorizing a formula again made Marco want to jump up and scream. Working with Mr. Pikake was turning out to be more wonderful than he could have ever thought. He saw stacks and stacks of notecards, some had formulas Marco had been tasked with memorizing, $A = \frac{1}{2}bh$ or $P = 2l + 2w$, others represented all the formulas he would amass over his next four years of math classes. He set the stack aflame and watched them disintegrate into ashes, the reflection of the fire burned in his light green eyes.

"Measurement is monumental!" Mr. Pikake began again forcing Marco to leave his daydream and rejoin his tutor in the real world,

---

[*] Seriously, no one understands formulas. Try it out. Take the Pythagorean Theorem for example. Ask someone if they know the formula. They can probably tell you $a^2 + b^2 = c^2$, but can they tell you why? Probably not.

"There are feet and fathoms, cubits and centigrade, dashes, drops, and so much more. It shall provide us a pulpit to practice substitution but," no pop. The professor lowered his voice to a whisper and slowed his words to a crawl, "we must be careful. Some of the battles won by Numberfolk – the battles where humans were overcome, were caused by none other than measurement mistakes, what we call conversion."

Marco was familiar with conversions, and he agreed they seemed like an unnecessary way to introduce mistakes. *How many ounces are in a gallon? Well, there are 8 ounces in a cup and 2 cups in a pint, there are 2 pints in a quart and 4 quarts in gallon. Why would anyone be confused?* He snickered. *Working backwards we have 2 pints in a quart and 4 quarts in a gallon so there are 8 pints in a gallon – right? Now 2 cups in a pint meaning 16 cups in a gallon and 8 ounces in a cup so that means* 16×8 *ounces in a gallon?* His mind was too tired to do the multiplication. "Why is it so important?" the boy was dying to know.

"Perfect probe!" Mr. Pikake shot out. "Before all this," he pointed his hands around the room, "one of the first things ever created was the Great Scale. Without measurement, we have no way to know what we have. Suppose I tell you that I have 4 logs and you have 6 rolls, which of us has more?"

The temptation to answer immediately pulled at Marco. Clearly 6 was more than 4, his mouth worked more quickly than his brain as, "I have more," escaped from his lips.

"Do you? What if 'logs' are yards and 'rolls' are feet. Are 4 yards truly less than 6 feet?" Marco knew his mistake before the professor finished the sentence. "No, 4 yards is 12 feet* which is more than my 6 feet."

"Completely correct! Such miscalculations have lasted through-out all of time. Columbus made such errors assuming all maps

---

* Since 1 yard = 3 feet, that means 4 yards is equivalent to $4 \times 3 = 12$ feet.

were in the Italian mile. What he thought was a short trip was in fact a journey around the world. Flash forward hundreds of years, NASA turned an orbiter into a lander as it crashed into the face of Mars. Why? Conversion. Not agreeing upon the language the Numberfolk are speaking can result in drastically different results."

NASA? Marco thought. *If they can't get it, how can I?*

"Here is an example." The professor continued, mistaking Marco's deflation with confusion. "A video game is nothing more than Numberfolk. Ones and zeros a human has coded to provide the patterns they must follow. You must then advert the obstacles provided. The same is true of our world. Take weather. Temperature, air density, wind speed – all Numberfolk. Once we understand their arrangement, we can pivot. We can't control the weather, but we can thwart it by building houses strong enough to withstand the winds and insulated enough to keep us warm. Every day we act to fight the Numberfolk – most people simply don't notice it."

If living in an animated world, a lightbulb would have appeared over Marco's head. His tutor's crazy was starting to make sense. All Mr. Pikake's talk about 'Numberfolk' was just his way of saying numbers are all around us. Crazy was only a matter of perspective. Cheering up, Marco leaned in.

"Now, if the weatherman told you it is going to be 30° today, what would you think?"

"I'd think I'd better grab my coat." Marco hated the cold. He couldn't wait for winter to end so he could be out on the baseball field.

"Very good… if the weatherman was talking in Fahrenheit! If he was measuring in Celsius, 30° is summer climate, and you would look pretty silly in your winter jacket!" Mr. Pikake laughed at his own joke. "When we understand the patterns, we can use them to adapt. Here is an expression that allows us to translate, like the enigma machine. His pen flowed across the stationary:

$$\frac{9}{5} \times C + 32.$$

"The Numberfolk hiding behind C is the temperature measured in Celsius. We can use this pattern as a tool to limit how much the Numberfolk affect us." Mr. Pikake pushed the paper towards Marco. "If 30 is hiding behind C, what do we see?" He winked.

Marco took the paper and pen and began to work it out. First, he used substitution to replace the C with 30.

$$\frac{9}{5} \times 30 + 32$$

Great. Fractions, he thought. *But this dosen't look too bad.* Marco knew that $\frac{9}{5} \times 30$ meant $\frac{9 \times 30}{5}$. He imagined the nine 30's all being ripped into five parts. As each one was pulled apart like rubber, they snapped back transforming into a 6. *That gives me nine 6's,*

$$= 9 \times \frac{30}{5} + 32$$
$$= 9 \times 6 + 32.$$

I can do this! He started to feel excited. *Nine times six is 54 and 54 plus 32 is...* He scribbled down his thoughts.

$$= 54 + 32$$
$$= 86.$$

"So, when someone says it is 30° Celsius, that means 86° to us? In Fahrenheit?"

"You've got it! Quite the difference, *ay!* Looks like you won't be needing that coat after all." The two chuckled.

"What about the other way? What if we know Fahrenheit and we want to find Celsius, do we do the same thing?"

"Always ahead." The professor smiled warmly. "In due time, lad. I will show you how to vanquish and how to obliterate, and then we will develop the f-word. We will swiftly sway back and forth, I promise." Mr. Pikake gathered the paper and pen, carefully placing the items back into his briefcase. "Our time has expired," he said with a twinge of sadness. "You have learned two key notions today. One," he held up his finger, "when given an expression such as $x + 7$, we can imagine what will happen by substituting different Numberfolk hiding behind the letter. This allows us to decode the pattern. For $x + 7$, as larger Numberfolk hide behind x, we would see the expression becoming larger and larger. And two." The next finger shot up, "Substitution may also work as a translation machine. Once we know the pattern the Numberfolk follow, we can translate between different units. Substitution is a very powerful asset for a hunter. So, keep this tool close at hand!"

The two exited the library and for a moment stood silent in the hall. Marco had the urge to hug his tutor. It wasn't just that Mr. Pikake was making math easier for him, it was that the professor had shown an interest in him, one that Peter never had. When the awkwardness had reached a peak, Mr. Pikake slightly bowed to Marco and softly said, "See you tomorrow." Marco returned the bow and the two parted in opposite ways down the hall.

When Marco opened the doors leading outside, the cold air slapped him in the face. He thought, *If 30° Celsius is a warm 86° Fahrenheit, it has to be closer to 2° Celsius out here.* He imagined little 2's swirling around him as he sprinted to his mother's car shivering.

To Marco's surprise, the car was quiet. Maggie had her entire head in a book. He tilted to read the title, *How to Win Friends and Influence People.* Rolling his eyes, he threw himself back into his seat.

"Baseball starts next week!" His mother exclaimed as she pulled out of the parking lot. Marco had completely forgotten. "You must be excited," she squealed.

He was. Marco looked forward to baseball season more than his birthday or the holidays. There was something about being on a team that made everything right in the world. The all too familiar sickly feeling crept into Marco's stomach. With the season starting, baseball would take over his afterschool time. When would he have time for tutoring?

Reading his mind, his mother chirped, "You'll have to do your tutoring on the weekends."

Maggie burst into laughter. "How fun! School *every* day."

Before, this news would have been devastating. Now, Marco no longer saw tutoring as taking away from his life, it was adding to it. Plus, this meant he'd have an excuse to get away from home on the weekends which was always a full 48 hours of Peter collapsed on the couch. He pulled out his phone and opened the weather app. Numbers littered the screen. There was the standard high and low temperature, but there was also a number representing air quality, one that determined the UV index, wind was measured in MPH and precipitation in inches, humidity was a percentage and pressure was measured in something called inHg.

Marco stared out the window. As the normal scenery passed by, not a single number was in sight. The crazy was starting to sink in as Marco adopted a new perspective. Although he couldn't see them, he had the strangest feeling that Numberfolk were hiding everywhere.

DECODING UNIT 1

While Marco's class is working on percents, ratios, and all that jazz, Mr. Pikake isn't so interested in these topics. That means, we don't really get to dive into these ideas much. Where better than a blank page to talk about them?

I have always had an issue with tipping at restaurants. I feel like, if I am required (put nicely, *highly encouraged*) to tip, why not just add it into the price? Turns out Argentina mostly agrees with me. They actually have a law saying you *can't* tip. In France, restaurants include the tip in the price. But in America, well, you'll need to do the math.

If you don't know, a tip is an additional amount of money you provide a waiter or other serviceperson. Generally, tipping is 10 - 20% of your total. How can you calculate this quickly?

Percents in multiples of 5 are straight forward. Since we use a base-10 system, to find 10% you need only to move the decimal over one place to the left. For example, 10% of $54 is $5.40. To compute 15% just take half of 10% (which is of course finding 5%) and add them together. Half of $5.40 is $2.70. That means a 15% tip on a $54 bill is $5.40 + $2.70 = $8.10. In the same way, you can compute any multiple of 5%. To find 35% just add 10% three times and then tack on the 5%.

What if you want to tip 17%? Well, you could move the decimal place over two spaces to the left to find 1% and then add. That is, 17% = 10% + 5% + 1% + 1%. In our example, 1% of $54 is $0.54, so 17% of $54 is $5.40 + $2.70 + $0.54 + $0.54.

Another quicky is to simply multiply. Since 17% of $54 is really $\frac{17}{100} \times 54$, you can just multiply 17×54 and then to divide by 100 you move the decimal over two places to the left. As $17 \times 54 = 918$, that means 17% of $54 is $9.18.

You are now ready to become an excellent tipper - you're welcome.

$$6 + (6 - 6) \times (6 + 6 \div 6)$$

$$\sqrt{36} \qquad\qquad \log_2 64$$

$$1 + 2 + 3$$

$$VI$$

$$3! \qquad\qquad 14 \bmod 8$$

$$6 + (6 - 6)\left(6 + \frac{6}{6}\right)$$

6

NOTATION

*"HE WHO CAN PROPERLY
DEFINE AND DIVIDE IS TO
BE CONSIDERED A GOD."*

-PLATO

The school day came and went without any excitement (unless you count lunch when Marco's classmate, Johnathan, spewed chocolate milk out of his nose). Marco soon found himself heading to the library. He was surprised to see Mr. Pikake waiting on the bench outside the big double doors.

"I thought a change of scenery might be nice." The man unfolded like a spider, towering over Marco as he stood.

"Where to?" Marco asked.

"I don't know about you, but I could use a snack." He motioned to the large sign that read NO FOOD OR DRINKS affixed to the library doors. Agreeing, the two started down the hallway in search of the cafeteria.

"I have to admit, son, I am impressed with how much you have grown in only a few days. You quickly mastered the ability to recognize Numberfolk disguises. You can easily see how someone like 48 might appear in many forms, such as four 12's, *and* you are able to continue to rip him apart to truly know what he is made of."

"The four is two 2's and the twelve is four and three, so two more 2's and a 3!" Marco chimed in˚.

"When we met, I'd dare to say the letters *scared* you. But now, you are beginning to master these as well!"

"Numberfolk masks can't fool me!" Marco said with pride. "I'll find whose hiding, snuff 'em out!" He flicked his wrist, fencing with an invisible force.

Mr. Pikake placed his hand on Marco's shoulder, a warm smile took over his entire face, "That you will, lad. That you will."

˚ The prime factorization of 48 can be expressed as $4 \times 12 = (2 \times 2) \times (4 \times 3) = 2 \times 2 \times 2 \times 2 \times 3$.

Together, they pushed open the cafeteria doors. The large linoleum tiled room lay empty except for the noise of the employees clashing and clanging plates and pans from a hidden kitchen filling the air. Mr. Pikake slid crumbled dollar bills into the vending machine purchasing himself and his student sodas and chips before they sat together at a nearby table.

"I have two goals for us today. One." His voice boomed freely without the eyes of Head Librarian Sabrina watching over them. "Notation. While a horribly dry subject, it will help you on both your examinations, and your ability to think quickly in battle and identify any Numberfolk foes."

Marco nodded as if to say 'check' with his head.

"And two – decoding!"

"Decoding what?" The words left Marco's mouth before his brain had any time to process them.

"Decoding situations to comprehend the battle that lay before us." The professor didn't look at Marco as he responded.

Reading between the lines, Marco groaned. He knew his tutor had just found a fancy way to say *word problems*.

"I know, I know. But hear me out. How often are you walking down the street and a Numberfolk expression just pops up in front of you?" Marco appreciated the imagery. He thought about riding his skateboard down the sidewalk when suddenly from behind the bushes $15 + x$ sprung to block his path. Marco tried to avoid crashing into it but failed, landing firmly on his butt. He smiled.

"Being able to decode a situation will be the first step in any battle. A good commander surveys the scene, pulls out the key information, and uses it to determine his plan of attack!" Mr. Pikake's arm shot up, assaulting the air with his finger.

"Okay. Okay. I'll give it a try," Marco reluctantly agreed.

Mr. Pikake grabbed the nearby menu sign and erased the daily special with his jacket's sleeve. He uncorked the attached blue marker and turned to Marco, "If I asked you to write two X's, how would you do it?" He passed the marker to his student like a runner handing over the baton.

Marco accepted and drew two X's on the menu board:

$$\times \times$$

Snatching the marker back, below Marco's X's Mr. Pikake added:

$$2x.$$

"Why not like this?" he asked.

A tightness clamped down on Marco's chest. This was the math of his nightmares.

Mr. Pikake leaned in, his nose almost touching Marco's ear and whispered, "Never fear the numbers, boy."

Trying to shake away the feeling, Marco rolled his shoulders before responding. "I guess... I guess I just don't know what that means." The $2x$ reminded Marco of one of his best friend's silly answers. Mrs. Sanders would ask 'How can we express four pumpkins?' Oliver's hand would shoot up, he'd approach the board and draw out:

$$4 \text{(ᗺ)}$$

The thought of Oliver made Marco's stomach turn. They hadn't talked since their lunch spar, and Marco found himself desperately missing his friend.

Mr. Pikake's long bony finger pointed to the number as he said, "Two," then slid his hand now motioning to the letter, "x". He did it again for emphasis, "Two. x." He let out a loud jolly laugh. "It means only that we have two x's!" He quickly added four strokes to the board to reveal:

$$x + x$$

$$= 2x.$$

$\frac{164}{b}$, 2 is hiding

"That's it?" Marco was shocked and embarrassed by the simplicity.

"That's it!" Mr. Pikake boomed.

"But, wait. That doesn't make any sense. Say 5 is hiding behind x. Then $5 + 5$, well that's 10, but two-five is 25. They aren't the same." Grabbing the marker back, he demonstrated for the professor:

$$5 + 5 = 10$$
$$25 \neq 10.$$

"Never neglecting to impress," Mr. Pikake grinned. "The whole thing is really quite silly, another way to confuse the masses." He scribbled on the board:

$$+ - \times \div$$

"Early on, we learn that these symbols tell us what the Numberfolk are doing, their non-covert operations." Marco nodded following along. "The plus tells us there is a duel, the $\times$ is proliferation, and so forth."

This was the second time Marco remembered Mr. Pikake mentioning Numberfolk duels. The word alone was candy to his brain. He loved the magical battles of video games and if numbers were dueling, he needed to know about it. He couldn't wait for his tutor to tell him more about this, but he sensed today wasn't the day.

"The problem is, no one can tell the difference between an $\times$ that tells us multiplication is occurring and an x telling us a Numberfolk is hiding behind their mask. So, the lazy mathematicians decided to just entirely eliminate the multiplication sign. When you see $2x$ that is really saying 2 times x or two x's. And as you discerningly detected, $x + x$ is 10 when 5 is hiding and, of course, 2 times 5 is also 10."

Marco wasn't convinced. "But how do they know two-five is two *times* five and not just twenty-five?" He was really digging in today. If he had to boil down his confusion in math it all started with notation. If he couldn't even understand the question, there was no hope at finding the answer.

"Surprisingly spot on! We use *juxtaposition*[°], placing the symbols next to each other with no sign – when there is a numbermask." He scribbled on the board:

$$2x \qquad 4k \qquad \frac{1}{2}m \qquad 0.56h$$

"You see? These expressions simply mean 'two times x', 'four times k', 'one-half times m', and 'zero point five six times h'." The professor continued to add to the board,

$$2 \times 5 = 2 \cdot 5 = 2(5) = (2)5.$$

"When there are Numberfolk and no masks, they replace the multiplication $\times$ with a dot or simply toss in parentheses."

Marco wanted desperately to take notes on what the professor was saying but resisted the urge.

"They've done away with the $\times$ to mean multiplication. They, too, have bid farewell to the division sign. Division is shown as a fraction."

The whole thing seemed entirely unnecessary to Marco. The familiar signs were like a baby blanket – a way to find comfort in even the most treacherous of situations. Now, they were being taken away, snatched from his hands. "And how are division and fractions the same?" He was just beginning to go through the stages of grief at the loss of the multiplication symbol when another familiar friend was ripped from his grasp.

[°] In short, we no longer use the multiplication symbol. When working with numbers, multiplication is represented as a dot such as $3 \cdot 10$ or parentheses $3(10)$. When letters are present, no symbol at all means multiplication. Something like $3x$ means 3 times x. If $x = 10$, then $3x = 3(10) = 30$.

"When we say one divided by two, it means we split what we have into two parts." Mr. Pikake leapt to his feet, his long legs took him to the vending machine across the room and back again in a mere six paces. He returned to the table with a candy bar. "I take what I have and split it into two parts. Then, we each get a half, a fraction." The tutor broke the candy in two and passed half to his student. He then wrote on the board:

$$\frac{1}{2}.$$

Marco's mind was all over the place and they hadn't even touched word problems. He began haphazardly nibbling on the chocolate bar to calm his nerves.

"This is how the Numberfolk hide from us. They obscure their identity – but, we cannot be fooled! For $4x$ just means the numbermask x is under an elargment spell of size 4."

Marco saw the x grow to be four times its size on the board. At first it scared Marco, intimidated him. Remembering the M&M's, he began to understand. "Like the candy, there were four x's – the number was hiding within them."

"Exactly! They are soldiers on a field and may call for backup by duplicating themselves," the professor boomed. "I hear you will be starting on the baseball team soon." Mr. Pikake sensed a change of pace was needed and Marco appreciated it.

"Yeah! I play first base, which everyone knows is the best. Way more runners come to first base than the others. Well, if we are any good that is."

"I suppose that means we should practice!" Mr. Pikake bent down and wrestled with his briefcase. When he popped back up, he held two mitts and a ball. He tossed one to Marco and slid the other onto his left hand. His tall fingers forced the glove to sit too high, barely covering his palm.

They separated themselves with ten linoleum tiles between their positions. Mr. Pikake tossed the ball to Marco, revealing his severe lack of skill. Marco lunged and easily caught the sideways pitch. He returned the ball to the tutor. After a few minutes, they developed a good rhythm, throw-catch-throw-catch.

Marco felt the warm, calming hug of the predictable smack of the ball hitting his mitt. The white noise of a ballplayer relaxed every muscle in his body. As they continued, he was struck with the realization that he had only ever played catch with his sister, his mother, his friends. Never before had there been a father-figure in his life who was willing and reliable enough to catch a ball. Peter had always been *there*, existed, but never inclined to donate his time. Marco didn't imagine it would be much fun to play with Peter anyhow. He could see Peter hurling insults as much as he hurled the balls.

"You see $5c$, Marco." Mr. Pikake tossed. "If 4 is hiding behind c what do we have?"

Marco caught the ball and threw it back to the tutor. "$5c$ means 5 times c, so 5×4. We have 20."

"Perfection!" In his excitement the professor nearly missed, he fumbled trying to grasp the baseball. They continued, shortening their expressions and regaining their rhythm. Mr. Pikake would hurl the ball and as it floated midair a second projectile came flying at Marco – a question. The student caught the ball and pitched it back along with the answer to his tutor.

Throw. "$14a$, 2 is hiding."

Catch. "We see 28."

Throw. "$4n$, 6 is hiding."

Catch. "We see 24."

Throw. "$100k$, 8 is hiding."

Catch. "We see 800."

Throw. "$50q, \frac{1}{2}$ is hiding."

Catch. "*Um.* That's 50 times $\frac{1}{2}$. So 50 divided by 2." Marco threw the ball back to Mr. Pikake. "We see 25."

Catch. "Wonderful! Let us increase our intricacy!"

Throw. "$\frac{1}{2}a$, 100 is hiding."

Catch. "We see 50." Marco was really getting the hang of things now.

Throw. "y over 2, 12 is hiding."

Marco caught the ball and hesitated. "y over 2. Do you mean that y is divided by two?" Mr. Pikake leaned to the side and grabbed the menu board. He struggled to grip it with the poorly fitted glove on his hand. Balancing the board on his knee, he scribbled:

$$\frac{1}{2}y = \frac{y}{2} = y \div 2,$$

before holding the board up for Marco to see, teetering on one foot with his leg in the air.

Having recently completed lessons on proportions, Marco dug through his memory. *One-half of twelve is the same as one-half times twelve which is twelve divided by two. Why are there so many ways to say the same thing?* He wondered. *This is what he means when he says the Numberfolk are trying to deceive us.* Marco pitched the ball back to the professor, "We see 6."

Throw. "Wonderful! How about x over 9, 54 is hiding."

Catch. This required speedy computation, but Marco was up for the challenge. The tutor was asking him to divide 54 by 9. "We see...6, again!" he laughed.

Mr. Pikake caught Marco's final pitch. He jogged towards the student in two leaps and raised his hand in the universal sign for high-five. Marco had to jump, but successfully smacked his palm against the tutor's. "You are no longer a novice of notation!" The professor shrieked. "In battle, you will need to survey the scene – simplify it." Rubbing his sleeve on the menu board, Mr. Pikake jotted:

$$\frac{1}{2}(10 + 8) - 3(2 + 1).$$

"Do you see, Marco? This looks *ugly*. It looks complex. But," the pop was like a bubble of gum had just exploded, "it is only a single Numberfolk trying to hide from us. Can you find it?"

Parentheses first. Marco reminded himself. He calculated $10 + 8$ and $2 + 1$ and slowly replaced the sums on the board:

$$\frac{1}{2}(18) - 3(3).$$

The one-half 18 really means one-half times 18 or 18 divided by 2 which is nine. Marco was considering each step slowly and methodically. He imagined himself a cheetah stalking its prey. *And the three three is saying three times three and that is nine, too*:

$$9 - 9.$$

"Seriously?!" Marco exclaimed. "All that for zero?!"

"Yes! You see! A Numberfolk as common as zero can disguise themselves well beyond recognition! But you hunted them Marco, you revealed their true self." The professor was giddy. It caught on and Marco's rollercoaster felt as if it was bouncing up and down on the tracks. "Let us maintain the momentum. Next, we will add in the numbermasks." With a few slashes of a marker Mr. Pikake revealed:

$$\frac{1}{2}(4x + 2) + 7(x - 3).$$

The track disappeared midjump and suddenly Marco was free falling. Like the multiplication sign, his confidence was snatched away, and fear gripped his body. Marco's good feeling was gone.

$$\frac{352}{x}, 4 \text{ is hiding}$$

He looked to the parentheses first but had no idea how to add four x's to two or how to take three away from an x. With a sharp tug, the rollercoaster landed and started pulling him down.

NO! Marco thought. His hands flew out on each side, grabbing the tracks to stop the inevitable dip. He was not going to let his emotions win this time. Only a few days ago, the sight of a letter in math was terrifying. Now, he was answering questions in the time it took to pitch a baseball. He didn't care that he was 'not a math kid', he could do this. He took a deep breath and let go. To his surprise, his emotional rollercoaster didn't plunge to the ground. Letting the tranquility wash over him, Marco looked to his tutor and in a firm and confident voice he said, "I need help. I don't understand the question." The emotions were no longer in control, the math was no longer in control, and if Marco could help it, the Numberfolk were no longer in control. Marco was in control.

Mr. Pikake looked down at him. For a moment, he thought his tutor had somehow seen his internal conflict. Mr. Pikake was staring at him in the way he imagined a father would stare at his son: kind, loving, but also bursting with pride.

"What a wonderful realization, Marco!" Mr. Pikake boomed. "Let's look deeper. If there were no numbermasks, what would you do?"

"Parentheses are first. But I can't add the numbermasks. What is $4x + 2$?"

"Very good! This will be our subsequent session. You are correct, you cannot simply combine these terms together. Without knowing who is hiding behind x, there is no way to simplify the expression. Move on, what would you do next?"

Instantly, Marco knew what he was being asked to do. *Distribute*. This rule was one that Marco had initially despised. The reason was very simple. He had finally gotten the hang of order of operations when everything was flipped on its head. We do what is in the parentheses first, then move from left to right doing all the multiplication and division before ending by moving left to right doing all of the addition and subtraction. Then one day, Mrs. Sanders wrote on the board:

$$8 \times (3 + 6).$$

Marco completed the parentheses first, $3 + 6 = 9$, and then went on to multiply 8×9 to find 72 when his teacher said, "When we have a problem like this, we can multiply *before* equating what is within the parentheses." He had no idea what to write in his notebook. MDPEAS? MDPEMDAS? Oliver had peaked at what Marco was writing and slipped him a note that had a doodle of a green pea with a stethoscope that was labeled 'Dr. Peas'. Five minutes later, Marco got the joke (doctors are often 'MDs' and it was a rift off Marco's MD PEAS) and burst into laughter which landed both boys in lunch detention that day.

Nowadays, he didn't hate the distributive property as much as before because Maggie had taken the time to explain the idea to him. Once it made sense and wasn't just another ridiculous rule, he began to warm up to the idea.

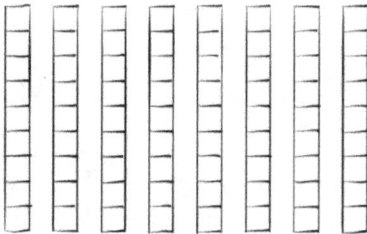

Maggie explained it like this, "You did it right, $8 \times (3 + 6) = 8 \times 9$." She drew out eight rectangles and then split each of them into nine pieces. "So altogether, you have 72 pieces."

Then she shaded in three of the pieces on each rectangle. "See, eight groups of 9 is just the same as having eight groups of 3 *and* eight groups of 6 put together."

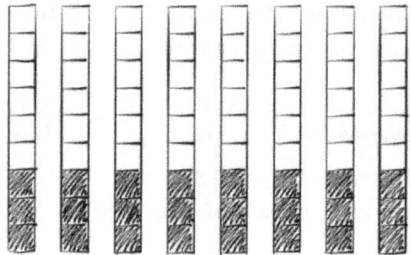

$18x$, 5 is hiding

After giving her brother time to digest, she concluded by saying, "So $8 \times (3 + 6)$ is the same as $(8 \times 3) + (8 \times 6)$. My teacher said that it doesn't really matter until you do letter math. Before then, you can always do the parentheses first, it is just a trick that can make things easier. I use it all the time to quickly calculate in my head. If I need to multiply 8×42, I think about it like $8 \times (40 + 2)$. Then 8×40 is easy, that is 320 and 8×2 is 16. That way I know that $8 \times 42 = 8 \times (40 + 2) = (8 \times 40) + (8 \times 2) = 320 + 16 = 336$."

Marco stood in awe at his sister's arithmetic gymnastics. He would have needed paper and pencil to stack the numbers and multiply them out, he couldn't even imagine being able to do all that mentally. At least he walked away understanding the distributive property and was feeling particularly thankful for her at that moment since he wouldn't even know where to begin if he had to add $4x + 2$ first.

"I have to multiply to distribute. The one-half is multiplying the $4x + 2$ and 7 is multiplying the $x - 3$."

"Ingenious!" Mr. Pikake yelled as he sprung to his feet. "Parentheses are nothing more than Numberfolk shelters. They are huddled together, working together, to thwart us. When you see the $\frac{1}{2}$ or the 7 outside, that is a curse cast upon the entire house, every term inside. In this case, shrinking or magnifying them all. The curse distributes itself to everyone hiding within."

Marco saw it.

$$\frac{1}{2}(4x + 2) + 7(x - 3).$$

He saw the $4x$ and the 2 hiding together in an abandoned mine, or was it a 4, an x, and a 2? A great wizard flashed their wand over the mine, tearing everyone inside in half. "So, we see $4x$ but the spell is really splitting that in half, does that make half of $4x$ or half of both four and x?" Marco questioned.

7x, 13 is hiding

"Recall, such a spell impacts each *term* inside these parentheses homes. For $4x$ is a single term, a single Numberfolk. That is $\frac{1}{2}(4x)$ has only one person inside, only one creature to impact. While $\frac{1}{2}(4x + 2)$ shows us two terms, two creatures."

"I see it. If I have four whatever's split in half, it would be two whatever's. And it is also hitting the two, and half of two is one!"

$$\frac{1}{2}(4x + 2) + 7(x - 3) = 2x + 1 + 7(x - 3).$$

"Yes. Yes. Keep going," Mr. Pikake urged the boy on.

"*Umm.* Okay. The seven is enlarging everything inside so the three is really seven times three or 21. The x, well that is seven times x, so just $7x$?"

"Perfection!" The professor shrieked, pointing to the board. "I was confident you would succeed." He made a fist like he had snatched the particles in the air before shaking it above his head.

$$= 2x + 1 + 7x - 21.$$

"Tomorrow we will tackle this more. For now, you have learned the art of substitution and the power of simplification. Simplification allows you to gain a clear picture of the battlefield, better understand the players. Then, you will be ready to dominate!" As he spoke, his words became louder, more powerful. By the time he said 'dominate', it was a bellowing scream that echoed in the empty lunchroom. Lowering his voice he continued, "Let us conclude with decoding, another powerful skill."

Marco nodded enthusiastically, he felt ready for anything.

"Suppose we wish to consider tickets for your baseball games. You would need one for your sister, one for your mother, and one for your father."

"Stepfather," Marco spit out, "and Peter doesn't come to my games."

"Fair point." Sensing this was not the time to begin picking into the boy's father issues, he moved on. "I would need only a single ticket, for myself, another family would need perhaps five tickets, or maybe two. The key here is we don't know how many tickets any person will purchase. It is hiding."

"You are saying that since we don't know how many tickets someone will buy, it needs a numbermask, like *t* for tickets."

"Precisely! And if tickets are $3 a piece, how would we decode this situation into an expression we can work with?"

"*Umm*. Okay. Tickets are $3, so one ticket is just $3, but two would be $6, and three, $9. You multiply the number of tickets by 3."

Mr. Pikake handed Marco the menu board. Marco scribbled:

$$3t.$$

"You got it!" the professor yelled.

Marco felt like he had hit the game-winning run, his teammates surrounded him and lifted him towards the sky. The crowd chanted his name.

"Now. There are 14 Numberfolk around us, we wish to split them apart to make them more manageable, but we are not yet sure the number of groups we will need to use. How can we express this?"

This seemed hard. Marco felt confident he would need a numbermask, but he wasn't sure for what. *Who is hiding?* he thought. He repeated the statement in his head. It came to him. The number of groups, we don't know that. "I think I need a numbermask, let's call it *g*, that is hiding the number of groups."

"Splendid. And what else?" The professor's eyes were sparkling with anticipation.

"*Umm*. Well, if there are 14 and we want to split them – divide them – into g groups, that would be $\frac{14}{g}$?"

"Amazing, boy! Here is another. Often, will we need to round up the Numberfolk. Suppose we have some twenties and three tens. How can we write this?"

Three tens is 3×10. Marco scolded himself for using the multiplication crutch in his imagination and quickly corrected it to 3(10). *We have some twenties, which means we don't know how many, so that is a mask. I will make it t for twenty. So that is* 20t. *Altogether we have...* He shouted out, "20t + 3(10)"!

"Never one to disappoint!" Like a blooming flower, a large smile spread across his face. "I shall convey a challenge! There are three Numberfolk, each is one more than the previous. Now the first has been enlarged by eight, the second has been torn in half, the third, is only himself. What is their sum?"

As each word left his tutor's mouth, Marco was thrown in a new direction, his rollercoaster travelling through a jungle of sharp turns jutting him from side-to-side. He didn't know the first Numberfolk, much less the second or third. *Do I need multiple masks?* Marco wondered. "I guess. *Umm*. The first one is n?"

"And if the first one is n, what is the second?"

"m?" Marco guessed.

"No, no." Mr. Pikake said firmly. "Don't make the battlefield more complex. The goal of the hunter is to simplify the situation. If the first is n and the second is one more, the second is..." he paused to let Marco continue.

"$n + 1$?" Another guess, another jerk to the side.

"Yes! If each is one more than he who came before, the first is n, the second is $n + 1$, and the third...what is it, boy?"

Marco needed a new approach instead of guessing. Although successful, it was making him feel sick. *One more than* $n + 1$ *is* $n +$

1 + 1, *which is* n + 2. He wanted to make sure before he presented it to Mr. Pikake. Deciding to use his new power, substitution, Marco imagined the first Numberfolk was 5. He had become fond of 5 for some reason. Every time he imagined it, the number was smiling and giggling. *If the first number is* 5, *the next would be* 6, *and the last* 7. *That makes sense because* 5 + 2 *is* 7. Confidence took hold and Marco clearly stated "n + 2."

"You are halfway there!" Mr. Pikake started dancing around the room, his long skinny legs bending reminded Marco of straws. "The first is enlarged by eight, the second torn in half, and the third is just himself!"

Beginning to write, Marco knew that an enlargement by eight was multiplication, so that was $8n$. The second number was cut in half, was that $\frac{1}{2}n + 1$ or $\frac{1}{2}(n + 1)$? He returned to his example. Since n was 5, the next would be 6 and cut in half would make it 3. Quickly computing, he found that $\frac{1}{2}n + 1$ would end up as $\frac{5}{2} + 1$ when he substituted the n with a 5, which didn't end up as the 3 it should be. Trying again with parentheses, he found $\frac{1}{2}(n + 1)$ was what he needed.

$$\frac{1}{2}(5 + 1) = \frac{1}{2}(6) = 3. \checkmark$$

He decided it made sense since the whole number, all of $n + 1$, was being cut in half, not just part of it. He saw the magician casting his suppression spell on the parentheses house as he raced to scribble down his thoughts before he forgot.

$$8n + \frac{1}{2}(n + 1) + n + 2.$$

"And there you have it!" Mr. Pikake grinned from ear-to-ear. "You have now gained the ability to both understand the written language we use to hunt and can decode situations and express them as well!"

Staring at the ugly expression, Marco couldn't help but feel incomplete. Sure, he had decoded, or translated, the situation into this new language, but the complex mix of numbers and letter-masks looked at him with disdain. "Is that it?"

With a hearty laugh the professor shared, "Is that it? Look at your power, Marco, your abilities! I will soon teach you the art of simplification. We will turn this offensive gang into a defenseless child. Using an equation, we will hunt it and locate the Numberfolk that is hiding, trying to deceive us. But," a loud pop rang throughout the room, "do not underestimate what you have already done. Translating the world into Numberfolk talk is precisely how you will conquer. It is the first step in your campaign to rule."

Marco, for maybe the first time in his life, felt tall. There was a bond with Mr. Pikake that he had never experienced with anyone before. The professor was offering Marco power. He wasn't the short little first baseman, he was a general preparing to command an army. Not only that, it was a secret world that only they shared. Not his sister or his friends or anyone else had what Marco had. He was special. Not sure how to express this, he opened up sharing his mistake.

"I almost thought it was one-half $n + 1$ instead of one-half," he paused "$n + 1$." They both burst into laughter realizing Marco had said the same thing twice although intending very different things.

"I know what you mean." The professor chuckled, "You almost applied the one-half to only the n rather than both the n and the 1." Marco nodded cheerily. "Classic slant mistake."

Marco thought he misheard his tutor. What on Earth was a slant? "A ꜱlant?"

"Yes, yes. A slant is what we call those who don't understand the power of the Numberfolk."

Self-doubt started coursing through Marco. It originated in the pit of his stomach and had already taken over his chest and his shoulders. *Am I a slant?* He wondered.

"You are *not* a slant. *You* are an aspiring master, and you continue to astound me." The professor read Marco's body language and quickly swooped in to build the boy up. "Time for one more?"

Determined to make his tutor proud, Marco raised his chin, "Bring it on."

"Alright," he grinned as if thinking of a diabolical question to pose. Marco was sure it couldn't get any worse than the last one. "Your baseball coach can buy bats in bulk for forty dollars each and gloves for twenty dollars each. Express this."

Marco was surprised by the question. It seemed much more practical than anything his tutor had previously offered.

Who is hiding? What don't I know? Feeling the mathematical momentum rising, he saw it. He needed a numbermask to represent how many the coach bought. He knew the price, but not the quantity. He wrote:

$$40b + 20b.$$

"Interesting." Mr. Pikake stroked his chin. "Only one Numberfolk can hide behind the mask b. That is to say, one b cannot hide a 2 and another a 3. What were to happen if your coach needed a different number of bats than he does gloves?"

Marco understood his mistake but wasn't sure how to fix it. He looked up at his tutor who bent and whispered, "You need another mask!"

Erasing the second letter with his sleeve, Marco changed the expression.

$$40b + 20g$$

"Well done, son. Well done." As the words filled the air, a bomb exploded inside Marco's chest. Hearing Mr. Pikake call him 'son' felt good, too good. More than he was willing to admit.

As they gathered their things to leave, Marco tried to make conversation by blurting out, "It seemed like we are in control of that last one! We decided how many to buy, it wasn't the Numberfolk."

A sad expression grew on Mr. Pikake's face. He knelt down to look Marco in the eyes, "You don't know how much I wish that were true. Numberfolk control inflation, that's how much money is worth. They litter the stock market boards and set the price of items. They even dictate how much a person is worth per hour of the day. We call this *Economics* and the Numberfolk dominate there, too." He slowly stood. "You are getting there, Marco. You are asking the right questions. You are starting to see the Numberfolk for what they really are – tyrants. Know that you are not alone in battle. I will be here, by your side."

They walked in silence until they reached the library when with a wave and a smile, they parted. Alone, Marco stared down the hall that led to the parking lot where his mother and Maggie would be waiting. Right before the exit, a second hallway perpendicular to his jutted out to the left. Marco could have sworn he saw someone peak out from behind the corner. He stopped. Keeping his eyes fixed on the place the two walls met, he carefully took a small step forward.

Suddenly, a thud came from behind. Marco whipped around to see 15 jumping off the clock that hung on the wall. He remembered 15. The 1 was big, angry, and marching towards him. Like before, 5 was giggling but it wasn't a friendly, playful laugh. It was a crazy laugh...a laugh you'd expect from someone in a straitjacket locked in an insane asylum. The 5 was being dragged

$\frac{x}{2}$, 196 is hiding

down the hall by the link that attached her to the 1's hip. Marco started running backwards, scared to take his eyes off the Numberfolk. Where was Mr. Pikake? He promised Marco wouldn't have to battle them alone.

A strong force hit Marco from the side. Stunned, he tumbled to the ground. His eyes shot back up to where 15 had stood, he saw nothing but an empty hallway. Marco turned ready to attack what had thrown him down.

"Watch where you're going, Marco. You ran right into me." His best friend Liam lay on the floor beside him. Liam crawled to his feet and brushed himself off before offering Marco a hand. "Whoa! You look like you've seen a ghost. Tell me *everything!*"

Stunned, Marco sat still for a moment before accepting Liam's hand and allowing his friend to help him up. "I-I must've been imagining things." Marco could barely get the words out.

"No dude. Ghosts are real. I have documented at least five myself."

The word 'five' made Marco jump. He walked with Liam to the parking lot, looking over his shoulder every few seconds to make sure the numbers were really gone.

DECODING UNIT 2

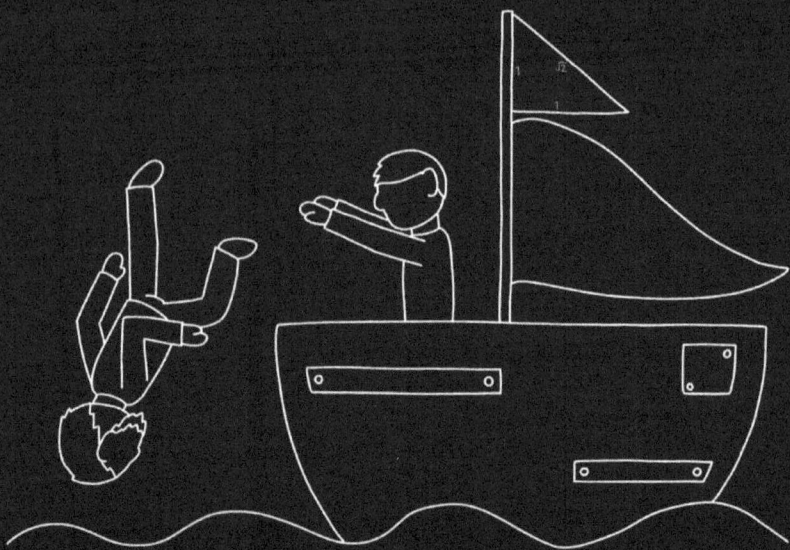

$(x + 1) - (-6 + x)$ $\qquad$ $10\left(\dfrac{1}{2} + \dfrac{1}{5}\right)$

$\sqrt{49}$ $\qquad$ $\dfrac{119}{17}$ $\qquad$ $\sqrt{3^2} + \sqrt{4^2}$

$2 + 5$ $\qquad$ $(2^2 - 1)^2 - 2$

7

MARCO THE GREAT

*"IF I HAVE BEEN ABLE TO
SEE FURTHER, IT WAS ONLY
BECAUSE I STOOD ON THE
SHOULDERS OF GIANTS."*

-SIR ISAAC NEWTON

This had been the slowest week in history. It was finally Friday, and the weekend was in sight! Marco was looking forward to some long overdue video game time. In fact, he planned to do nothing this weekend *except* play *Zombies: The Undead Tale.*

Still shaken from his encounter with 15, Marco kept a close eye on every number he came across. Luckily, none of them tried to attack. When the final bell rang, Marco sprinted to the library and was relieved when he found Mr. Pikake back at their table.

Out of breath, he began describing his Numberfolk encounter to the professor. His words flew out of his mouth at impressive speeds.

"Son. Son. Calm down. It's okay." Mr. Pikake lovingly squeezed his student's hand. "You are awake now. That is a good thing. You are beginning to see what few others are aware of. You recognize that Numberfolk are everywhere." He squeezed tighter and looked Marco straight in his eyes, "They know you are working with me, with SAN, and they will try to stop you. They control everything, and they are everywhere, but we know how to protect ourselves. You will become a great master. One day, you will protect not only yourself, you will have the power to protect many. I dare say – it's your destiny, Marco."

The tutor released the boy's hand and reached deep into his briefcase. Bending in half, Mr. Pikake's head disappeared beneath the table. When he returned, his long bony fingers were wrapped around a thick book. Carefully, he placed the tome in front of the student. Marco examined the cover. The book was the oldest he had ever seen. The pages were constructed of a thick parchment. Unlike the thousands of books that surrounded the two on the library shelves, its pages were not smooth – they were jagged like teeth. It was bound with what seemed to be animal skin, soft and smooth yet uneven. Marco opened the text to see it was entirely handwritten, not printed from a press. Words took over the pages with long strokes made with deep black ink.

$$8^2 + 5^2 + 3^2 + 2^2$$

The Society for the Abolishment of Numbers

A History

Volume 9

"Few have ever seen this book, Marco, much yet held it. This volume recounts the many masters who have come before us."

Marco began delicately turning the pages. It appeared to be a collection of biographies. He read the names aloud as he scanned the pages. Many he had never heard before, but he did recognize a few.

"Thales, Pythagoras, Zeno, Euclid," he advanced to find, "Sun Tzu, Brahmagupta, Al-Khwarizmi." Skipping farther ahead he read, "Napier, Mersenne, Descartes." The book naturally bent along the spine thrusting Marco forward about fifty pages, "Fermat, Pascal, Newton, Leibniz." Turning to his tutor he said, "I don't understand."

"Many men and women all saw, like you are beginning to, how powerful Numberfolk are. They all attempted to understand, and, in many ways, overcome them. Consider the followers of Pythagoras. In the early days, the SAN considered all Numberfolk to be commensurable. This is a fancy way of saying that all Numberfolk are rational beings. But all humans are not rational, and there are infinitely more Numberfolk than there are humans. For surely there must've been at least one that was *irrational*. Well, eventually a Pythagorean found one. It tore the community apart. Some say the man, Hippasus, who identified this Numberfolk was taken out to sea and thrown overboard. People were so scared of what this might mean. Understandably! Imagine a crazy untamed Numberfolk. They'd rather think the man a liar than believe his claims to be true."

$$10^2 + 1^2 + 1^2 + 1^2$$

A chill ran down Marco's spine as he thought of his brush with 15. The idea of an irrational number terrified him. "What does it mean to be irrational?" his voice quivered.

"*Ah*, excellent inquiry. A rational person thinks logically. It makes them predictable, reliable. Rational Numberfolk similarly behave *nicely* in a way. They are made up of two integers – a fraction – with one on top and one on the bottom. Irrational Numberfolk on the other hand cannot be created in this way. They are hypnotizers, they play a mesmerizing tune that cause digits to blindly follow them. They are hard to tame, hard to understand. You have probably heard of at least some of the most notorious irrationals. Pi (π) is one. Another is the square root of 2 ($\sqrt{2}$). We also have Euler's number, e. All of these contain decimals that continue forever and ever, never repeating. It makes them unruly because we never really know what they are going to do next unless we memorize their digits, their kidnapped victims. Even with that, we can only remember so many."

Marco imagined π in a straitjacket. It was drooling, its eyes crazy and wild. It thrashed back and forth like a rabid dog. He shuddered.

Mr. Pikake continued, "On the matter of irrationals, eventually, all members of the SAN, Saints we call them, had to accept what Hippasus had found. Irrational Numberfolk exist. It progressed our understanding, but the whole ordeal is an embarrassing stain on our history." The professor flipped the pages back to almost the beginning and pointed at the name scrawled across the top. "Another notable master was Euclid. Even in his time, few people understood the importance of the SAN. His work was vital as he set forth to prove to the world that every theorem, the traits we knew about the Numberfolk, was

$$6^2 + 6^2 + 4^2 + 4^2$$

truth. But oh, they dragged him through horrible trials. They disputed his work, said that he had assumed too much."

"What did he assume?" Marco was more than intrigued by these stories. The idea that a man was killed over numbers was fantastical, the fact that it was entirely true was unsettling.

"Euclid's idea was to start at the beginning. Using as little as possible, he attempted to show that all the ideas about geometry and Numberfolk had to be true. Oh, I can't wait for us to study his *Elements*, son. What Sun Tzu's *Art of War* is to a soldier, Euclid's *Elements* is to a number hunter. The controversy was all about one of the ideas Euclid assumed to be true." The professor held out his arms in front of him like a zombie. He then bent his elbows, so his forearms made an equal sign. "He supposed, in a way, that if two lines are such that the angles made with any line that crosses them are together not equivalent to two right angles, then the lines must eventually touch. In other words, the lines aren't parallel." He dropped his right elbow down to meet his left palm.

Marco didn't understand. This didn't seem like some horrible offense. "What's so bad about that?"

"Haha." The professor chuckled, "It is too much to assume. The statement is true in a small enough space, but Euclid needed to *prove* it was true, not assume it. This assumption caused much

$$8^2 + 6^2 + 2^2 + 1^2$$

debate around his entire anthology despite all the brilliance it contained."

"Small enough space?"

"I shall tell you the full tale when we study geometry. The geometry in schools is horribly misleading – they won't tell you everything. You see, schools today all utilize Euclid's geometry. While wonderful, it is the geometry of a small space. Consider the Earth, you know of longitude and latitude, yes?"

Marco nodded. He had studied geography in sixth grade. He knew the longitude and latitude lines were the ones that circled the Earth, like the equator. Although he always had trouble remembering which ones went up and down and which went side-to-side.

"We call the lines parallels. Next time you look at a globe, notice that the lines of longitude all meet at the North and South Poles."

"But why are they parallel then? I thought parallel meant they never meet?" He remembered parallel and perpendicular from when he learned of squares and other shapes*.

"Precisely! If the world was flat, the lines *would* never meet! Euclid's geometry is the geometry of a flat world, a world without curvature."

"You are saying they teach us flat world math? Why would they do this?" At first Marco felt puzzled, even outraged. Those feelings quickly subsided as he rationalized that most of what he

---

* Perpendicular lines are lines that meet at right angles to form L-shapes, like the corner of a table. It depends on who you talk to when it comes to parallel lines. Some say they never touch, meaning a line cannot be parallel to itself. Others claim parallels are lines that are the same distance apart at any two perpendicular points, meaning a line *can* be parallel to itself. We know – confusing.

$$6^2 + 6^2 + 5^2 + 3^2$$

had been taught had no connection to the real world or everyday life anyhow.

"That is where the small-enough comes in. To us, because we are so small, the world may as well be flat. Architects and construction workers use Euclid's geometry every day, because what they build is tiny compared to the size of the Earth. If they built something as big as an ocean, then they'd need to consider the curve of the Earth and that is where Euclid's assumption would cause them issues."

"I think I get it!" Marco had a feeling he would enjoy geometry, the idea of the math of building reminded him of the hours, days, he spent on *Minecraft*. "We can use Euclid's geometry because in a small space the curve of the Earth is so tiny, it is basically flat."

"Bravo! We call that *negligible*, small enough that humans feel comfortable ignoring it. You can test the idea by trying to draw a triangle on a soccer ball. You would have trouble using straight lines, but the smaller the triangle you draw, the easier it will become."

One thing was still bothering Marco, an itch in his brain he couldn't scratch. "If his assumption was wrong, and everything came from that assumption, isn't it all wrong?"

"Not at all, son. Have you ever known something that you can't prove?"

Marco's mind whipped back to last year. He had placed his phone on his desk before going downstairs to play video games with Liam and Oliver. When he returned to his room, his phone was gone. His mother was livid he'd lost something so expensive. He swore it wasn't lost – it had been stolen. She didn't seem to care. Marco was *sure* his kleptomaniac sister had snatched it. Maggie put on her puppy dog eyes and insisted to their mother she had done nothing. Eventually, it magically returned to his desk one day after school. "Yeah. Anytime my sister does something sneaky. I know it was her, but I can't prove it."

$$9^2 + 4^2 + 3^2 + 1^2$$

"Very good! And the same is true for many number-hunters. They detect a pattern they know is there, but they simply cannot prove it *must* be there. As time goes on and we learn more and more, we can often return to those unsolved problems with a new perspective. For Euclid, later Hilbert came in and improved on his original work. And the great thinker Gödel showed that there is no set of assumptions that will allow us to prove all truths." The professor was beginning to speak more quickly, more passionately, "But no one ever listens! They take what they want to take and leave the rest." He placed both palms on the table, gripping hard, "They turn our work into equations and examinations for school children. They make the world despise mathematics, keeping everyone in the dark! We are lucky we are relevant at all! Now, *now*, some are suggesting removing the study of algebra – the art of identifying numbermasks from the curriculum entirely. They push everyone into a cave and force them to see only shadows of the world. But you, son, you have broken free! You are beginning to step into the light and see the true nature of the world. You will see what causes those shadows. You will see the control of the Numberfolk!"

Marco had no words. Not even his vivid imagination could make sense of what lay before him. Before today, Mr. Pikake was just a crazy math teacher. He might go as far to say the professor was a bit *irrational*. As he looked over the book that lay between them, Marco saw things from a new point-of-view. His tutor was part of something bigger, something that stood the test of time, something that many famous names had built.

Mr. Pikake regained his composure, sitting up straight. "This is a lot to take in, I know. You have glimpsed past the shadows, Marco. You have started to see. You have greatness in you. Such greatness. One day, they will add your name to this book. This is something I know but cannot yet prove. You are remarkable, son. Simply remarkable."

Marco felt empty, the way your stomach feels when in need of a meal. Painful. Like something was missing, like something was

$$7^2 + 5^2 + 5^2 + 3^2$$

needed. No one had ever believed in Marco. His stepfather tortured him, his sister belittled him. His mother thought the world of him, but that is just what mothers do. As Marco stared into his tutor's deep blue eyes he saw hope, he saw love, he saw truth. Mr. Pikake believed what he was saying, he believed in Marco. He saw something no one had seen before. He saw Marco as something more than he was. It was a feast laid before a starving child and Marco ate. He devoured every last bite until he was no longer empty but filled with purpose.

"I need you, son. The world needs you. The SAN needs you. You have the ability to change things for the better."

Why can't I become a great master? Look how far I have come already! The tutor believed in Marco and in turn, Marco believed in himself.

"I'll do it," he finally replied. "I'll learn to be a master. I will make you proud." As the words left his mouth, a stabbing pain shot through his chest. He knew this meant trouble. Trouble with Peter, trouble with Oliver. But Marco didn't care, he pushed the thoughts away and placed all his focus on the professor's words. For the first time, he was needed, he was important. He wasn't going to let anyone ruin that.

If Mr. Pikake's smile stretched from ear-to-ear before, it doubled now, entirely overcoming his face. "You already have, son," he said softly.

As they gathered their things to leave, Mr. Pikake bent down and whispered, "I didn't even tell you my favorite story!"

Excited, Marco bit, "Tell me!"

"Well, one of the most famous unsolved problems was from a master named Fermat. He wrote his findings in the margins of a book with the words 'I have discovered a truly remarkable proof which this margin is too small to contain'."

"That's it? Did he write it somewhere else?"

$$8^2 + 5^2 + 4^2 + 2^2$$

"No! It wasn't until 350 years later that a proof was found. And even better, it was proven with ideas Fermat could not have possibly known in his time."

"Do you think he was lying?"

"Ha! It is one of the great unsolved mysteries. Did he have a proof? If so, what was it? We might never know." Mr. Pikake's eyes glimmered with intrigue.

On the car ride home as Maggie cheerily chatted about her day, Marco pulled out his notebook and turned it to the side. In the margins of the page he jotted, *I have found something remarkable which this margin is too small to contain.* Looking at it, he smiled before adding a signature – *Marco the Great.*

❖ ❖ ❖

A fire blazed heating the small room. Mr. Pikake paced. There was only space for two large steps before turning and taking two more. In front of the hearth sat two velvet armchairs atop a worn circular rug. Behind them, a gigantic chalkboard stood riddled with numbers and symbols. The ghosts of what had been erased still haunting its dark green face.

"He knows of the numbermasks. He understands that Numberfolk hide behind letters to thwart us. He is skilled in substitution and is beginning to learn how to represent patterns through expressions. There is just so much to learn. He has excelled in a week's time, but I worry the young boy can only take so much. He knows not of the first number wars, or the Vinculum Games. Certainly, he is not ready to know more of the irrationals."

A loud screech filled the room. From behind the chalkboard a stocky man appeared in the shadows, dragging a piece of chalk coarsely across the board. He wore all black, a three-piece suit, and a top hat as if he had just left a lavish event.

"Calm yourself, Blaise. There is time. Continue to teach him the ways. Inform us of any developments."

$$7^2 + 6^2 + 4^2 + 3^2$$

I have found something remarkable which this margin is too small to contain. –Marco the Great

As quickly as the man appeared, he vanished. Mr. Pikake sat. He grabbed his hand trying to stop the trembling that had taken over. The weight of the world lay on his shoulders. It was his responsibility to shape Marco into everything he needed to be. But Marco was just a child. It was so much, too much, to place on the boy.

Yet, none of this was what was bothering Mr. Pikake. It was the way Marco looked up to him, the way Marco trusted him. It was the laughing. It was playing catch and splitting a chocolate bar. It was dancing through the library stacks. It was love. In no time at all, Marco had become like a son to him. It was this feeling that shook the professor to his core.

$$6^2 + 5^2 + 5^2 + 5^2$$

f_6

$2^2 + 2^2$

$(0.2)(40)$

$(x-1)(x+1) : x = 3$

2^3

$4!!$

$\dfrac{4}{\frac{1}{2}}$

$16\sin\left(\dfrac{\pi}{6}\right)$

ZOMBIES AND BASEBALL

"MATHEMATICS IS A PLACE
WHERE YOU CAN DO THINGS
WHICH YOU CAN'T DO IN THE
REAL WORLD."

-MARCUS DU SAUTOY

Bright rays shone in Marco's window. He rolled over and pulled the blanket across his head. The tranquility of waking up naturally was a luxury that came only on the weekends. On school days, Marco was rudely forced awake, the sound of his alarm screaming 'Wake up! Wake up! Wake up!' before the sun even had a chance to rise.

Marco was in a daze. Had yesterday really happened? Or had he imagined it? As he twisted out of bed and slipped on sweatpants and a t-shirt, his mind stayed firmly on Mr. Pikake. The memory was too clear for it to have been a dream. Marco was back on his personal rollercoaster and in an interesting change of pace, it was no longer speeding through loops and whirls, or racing down devastating drops – it had switched direction completely. He was flying backwards being pulled and pushed in new ways. He recognized the feeling, it was regret.

Why had he so easily agreed to become part of this secret society? As much as he wanted his strength over numbers to translate to dominance over Peter, he knew better. What was he thinking? In truth, Marco knew exactly what had persuaded him. Mr. Pikake had shown an interest in him, he cared about Marco, and he needed Marco's help. And if Marco could help, he should. The rollercoaster was losing speed.

Maybe it wouldn't be that bad. Maybe Peter was all talk, and Marco could certainly mend things with Oliver. His insides didn't agree. With a sharp pull, Marco was hurdling forward now. A new feeling was taking over – fear. He remembered the terrifying encounter with 15 in the hall. He remembered how numbers are everywhere. He heard a rustling behind him, someone was there, he could feel it. He whipped around to see only his messy room, things piled everywhere.

As he carefully studied his surroundings, trying to convince himself he was alone, numbers began appearing all around him. His bottles of sports drinks all had numbers on the labels explaining the nutrition facts. There were numbers on his clothes, on his games. There were even ISBN numbers on every book he owned. He sprinted out of his room, down the stairs, and into the garage. Throwing anything that was in his path out of the way, Marco finally found what he was looking for – painter's tape.

In a dash, he made it back to his room. He began placing the thick green tape over every number he could find. He scanned the room again, not a number in sight. Relieved, he tumbled back onto his bed. In his panic, he had worn out all the adrenalin that was coursing through his blood and was now able to allow the calm to wash over him.

He let out a loud groan as he hugged his pillow and tightened his body into a ball. There was no stopping the runaway train he was on. He saw the Numberfolk now, he knew they were there, and he couldn't magically put the genie back in the bottle. He needed Mr. Pikake to teach him how to become a master, otherwise he would be completely helpless. He'd have to figure out both the numbers and Peter. He lay there for a moment, cursing his life and its complexities before being jerked out of his head to the sound of his bedroom door flying open.

"Good Morrrninggg!" Maggie screamed like a sportscaster announcing the starting lineup. She flew (yes, actually flew) across the room, taking a large leap and swan diving onto Marco's bed. While this early intrusion would normally get on his nerves, the sight of his sister was a welcomed distraction and a reminder that he still had someone in his corner. "I hate to say this. And if you ever tell anyone, I will deny it and leak your darkest deepest secrets…" She continued in a peppy but terrifying tone, "but…I missed you this week. I feel like you have been off in your own world more than usual."

Maggie was right. At family dinners, Marco had barely said a word (although there were not normally any breaks in his sister's constant jabbering for him to interject anyway). He'd had a lot on his mind, and even with his sister willing to be a friend, he wasn't ready to share with her. How could he? Peter was her real dad…

she'd end up taking his side or making excuses for him. Afterall, Peter only ever showed Marco his monster mask.

"Sorry, Bug. First week back to school has been tough. Plus, I am trying to not completely flunk out of math."

"How's that going? You know, working with the elderly?" She swayed back and forth on the bed laughing insanely.

He rolled his eyes and pushed her off before getting up and starting towards the door. "Mr. Pikake's really nice. Don't make fun of him like that."

The two locked eyes, each knowing what their sibling was thinking. A silent foghorn blew to indicate the start of the race and they were off. They shoved each other as they ran down the stairway. Marco grabbed the door frame as he turned the corner to give him the stability he needed to take the last three leaps to the computer desk. He slid into the oversized office chair. His butt sinking into the faux leather was enough to declare him the victor.

The family had a shared desktop computer in the first-floor office. It was a Saturday morning ritual between the two to determine who would get first use. "*Ahhhhh. Ahhhhh.* The crowd goes wild!" Marco shouted, boasting his win.

Maggie crossed her arms over-dramatically, "I really need to catch up on my podcasts! My followers need me!" she whined.

"Use the iPad, I got here first. It's Zombie time!" Marco didn't even look up, he was already opening the game and sending chats to Liam and Oliver to join in. Maggie stomped away. When he felt like it was safe to get up, Marco shut the office door then settled in with headphones and mic on.

"Where to boys?" Liam chirped in Marco's ear.

"The mountain," Marco suggested, "we have to get over it eventually. And we already got everything we can from this town."

"*Ug.* You would say that." Oliver burst in. His dissatisfaction with Marco rang out as clearly as his words.

"Marc's right. We gotta keep moving forward," Liam responded and with a reluctant grunt Oliver agreed.

The boys commanded their avatars to start the trek north to the mountain. The three walked in a phalanx formation, back-to-back, to ensure no surprise attacks. A few rogue zombies would pop-up here and there, nothing they couldn't handle. At one point, Marco swore that Oliver let one through to sneak up on him from behind. He decided not to say anything to avoid the inevitable argument that would follow.

They reached the base of the mountain and began their trek up. "Up there, on the right. See that ledge? Looks like a cave, let's explore it," Marco instructed. He took the lead as the three successfully traversed the terrain.

They arrived at the small rock clearing and looked around. A deep cave extended into the mountain, too dark to see inside. Oliver had picked up a flashlight in town. He went first, shining the small circle up and down the walls. Out of nowhere, arrows started flying towards the group. "I'm hit!" Liam screamed. Using their makeshift shields, they deflected the arrows inching deeper and deeper into the cave. They turned a corner and the cave expanded, revealing stalactites hanging from the ceiling and dozens of smaller caves projecting from the larger room.

"Stop! Stop!" An NPC approached them and the assault ceased. "Welcome Friends! We apologize for the security measures. Can't be too careful."

Now able to lower their weapons, the three explored their surroundings. They had happened on an entire civilization hidden within the mountain. They looked down to see a spiral staircase, held together with sticks and ropes, that led to an open area marketplace where people gathered trading goods and services. The far back end of the cave held long tables arranged perpendicular to a stage. Atop the stage was an enormous throne that sat empty.

"My name's Sam," the NPC continued. "You can think of me as the welcoming committee. You are welcome to stay for the night and enjoy our community to refresh yourselves before continuing on your journey."

"Eyes up!" Oliver commanded. "Something's fishy here."

"Yeah. We've never come across a full room of NPCs before, well non-zombie NPCs that is. Who knows what kind of trap this is. I took damage, I need to regain my health so we might as well look around."

The boys stayed together as they explored the marketplace. It was a gold mine of supplies. There was food, weapons, and armor being offered at every stall.

"Hey! This one's offering a health potion. I am going to try to trade with him," Liam said as his character engaged with the seller. Marco and Oliver continued forward together but the clear tension between the two made Marco feel alone. Then he saw it. The weapon of his dreams. He had wanted a crossbow ever since they announced its addition in the last update. But, he had never seen one anywhere in the game. Now, there it was, as if beckoning him. He split from Oliver and engaged with the seller.

"Interested in the crossbow, eh?" The NPC's words flashed across the screen. "I'll make you a deal. The crossbow is yours if you will agree to help my sister. The mayor here is an evil man. He has taken her and the other young children to a secret lair." Marco's view twisted to highlight one of the long hallways projecting from the main room. "Save her, and it is yours."

"Looks like a..." Marco started on the headphones before all three simultaneously shouted, "BOSS BATTLE!"

"I traded some gear for the health potion, so I'm ready," Liam added in. The three came together and started towards the path to the mayor. The hall was dark except for tiny lanterns that were spaced just enough to light their way.

"HELP US!" scrolled across the screen. To the left, they saw a jail cell where the children sat shackled. The three continued forward until they came to another opening. Much smaller than the marketplace, this cavity was only big enough for a player to take one good jump and end up on the opposite side of the room. The height of the cave – that was impressive. It soared up so much that the roof was out of sight. On the far right, a man sat with his back to the entrance.

"That's him!" Oliver yelled as he ran towards the character. With his bat, he landed a strong blow knocking the man from his chair. The mayor yelled a loud bellow that shook the caves, a stalactite fell from above almost crushing Marco. The mayor unfurled his body like a spider to reveal his unnatural height.

"Marc, you go left, I'll go right!" Liam yelled. Marco jumped to the left as instructed, he began hacking away at the mayor's leg. Gaining his attention, the mayor turned to face Marco. Marco's mouth dropped open. Frozen in place the character swiped Marco like a gnat. He flew into the wall and the screen went black. He could still hear Liam and Oliver, "What are you doing, Marc?!"

"Right, right, go right!"

"I can't! There isn't enough room in here to maneuver."

Marco sat staring at the black screen. *It couldn't be, could it?* The shock subsided just enough for Marco to move. He leaned in. Behind the layer of black he could still make out his teammate's battle with the mayor. The monster's back was to the screen and Marco struggled to get a glimpse of his face.

"I'm out," Liam cried. It was only a few moments later that the screen switched from black to red indicating they all had perished.

"What is with you, Marc?" Oliver shouted, "It was like you just let him take you."

"I-I thought I saw something. Got distracted," Marco replied.

The three regenerated back in the town at the base of the mountain. They started towards the cave. This time they were ready for the arrows and easily dodged them. Marco spammed the controls to get through Welcome Committee Sam's message and before long they were back in the mayor's lair. All three of them lasted longer the second time. Marco put up a good fight despite the fact that throughout the entire battle he was wrestling to get a good look at the character's face. Liam and Oliver kept their attacks tight, but the room was too small to do much. Marco tried and tried but couldn't manage to get a good view. The mayor's height didn't help the situation. His head stood high in the cave allowing it to be mostly obscured within the shadows.

The boys respawned three more times. Each time they lasted a little longer but were never victorious. Marco's mother was screaming for him to come eat lunch.

"Gotta run, guys. Lunch," Marco announced.

"Yeah, me two." Liam echoed. "Back tomorrow morning?"

The three agreed and signed off. At lunchtime, Marco received a lecture about the state of his room. His mother gave him an ultimatum: clean it or no video games for a week. He dragged himself up the stairs and collapsed on his floor. He had only planned to stay there a minute, but emotional exhaustion took over and he easily dozed off. When he awoke, he had no clue what time it was – he had taped over the numbers on his clock. The amount of light outside his window told him it was late afternoon, almost sunset.

He reluctantly began cleaning. Gathering all the trash first, he grabbed the empty containers that once held snacks and drinks and tossed them into the bin. He scoffed as he looked over the tape obstructing their calories, sugars, and the sort. It would have been a better idea to have just ditched them that morning instead of wasting his time covering numbers that were destined for the street anyway.

After taking the trash outside, he headed back upstairs. Using the smell test, he sorted his clothes into two piles: dirty and sort-of-clean. He threw the dirty pile into the washer and stuffed the sort-of-clean pile back into his closet, fighting with it to close the door. The entire time he spent cleaning, all Marco could think about was the video game character. He ran the game scenes back through his mind trying to remember exactly what he saw – or at least what he thought he saw.

The rest of the day sped by. Dinner was the Maggie show. She talked and talked about her podcasts and didn't seem to care that no one except their mother was listening. Peter stared Marco down the entire meal. Relief washed over him when he finally made it back to his room. It was much nicer without all the junk everywhere, but he'd never admit his mother was right.

He rushed to bed and tried to force himself to sleep. He wanted nothing more than to get back to the mayor. The game ran over and over in his mind until he was jostled awake by the sound of Maggie's voice. It was like she hadn't stopped talking the entire night. As the sun shone in the window, Marco pulled his pillow over his head to drown out her jabbering when it hit him. It was coming from *downstairs*! He sprung up and sprinted to the office to find Maggie already sitting at the computer. She flashed him a smug look of satisfaction without skipping a beat.

"Seriously?!" Marco blew up. "I am supposed to play with Liam and Oliver this morning!"

The two argued at their mother, who ultimately ruled in Maggie's favor. "You had the computer yesterday morning, Marco. It's your sister's turn." Maggie stuck out her tongue and ran back to the office. She was recording her weekly YouTube content.

If you heard it from Maggie, you'd think she was an internet sensation. "My followers depend on my unbiased commentary of the week's news." In truth, she had about 50 subscribers and a handful of views. Nevertheless, she religiously listened to her podcasts, taking notes so she could compile her take on the state of affairs on everything from what happened in Congress to the outcomes of her favorite reality shows that week.

Marco sent Liam a message saying he wouldn't be able to make it before moping over to the couch to flip through the television channels. Peter had been called out on some emergency extermination job which bought Marco time before he would be forced to confront him again.

"Baseball starts tomorrow," his mother said as she entered the room. Marco appreciated her attempt to cheer him up. "Do you have all your gear ready?"

He didn't. He wasn't sure how adults could possibly keep track of everything without the constant reminders from their mothers. He shook his head. It was thirty more minutes before he could dig up the motivation to move.

Marco threw open his closet door. The pile of clothes he had stuffed inside launched an avalanche as everything tumbled down onto the floor of his room. He dug through the wreckage. Reaching deep, he felt around for the vinyl bag. Feeling the cold slippery plastic, he tugged until the clothes monster was forced to release it. Inside was his bat, a few balls and two gloves stuffed right where he had left them. He picked through the rest of his stuff, making sure he got everything before cramming the mountain back inside. He sat on his bed and started taking inventory. All was present except his uniform. Marco had sported #13 for the last two years. If he didn't bring it back this year, he could easily be assigned a new number which was a sure way to jinx the entire season.

He crawled onto his stomach to search underneath his bed. Marco was a true believer that 'clean' meant 'not easily seen'. Any hidden area, the closet, beneath the bed, in the crack between his desk and his dresser, was riddled with junk Marco didn't even know he had. After unsuccessfully pulling out one thing, then another, and another, Marco finally located the last of his gear. He shook the shirts and pants out one at a time, attempting to release the wrinkles. The last thing he threw into his bag were his cleats. As he grabbed them, he noticed a small piece of paper in the bottom of his left shoe. Instantly, he knew just what it was.

For the last four years Marco had dressed up as a baseball player for Halloween. It was a go-to costume that didn't require much prep work. One year, Maggie was really into makeup and convinced him to go as a vampire pitcher to provide her with a fresh canvas to practice on. He ended up with a face full of paint. She started with a white base to sell the 'dead' look, which ended up making him look like a clown. Insisting it was just the first layer, she proceeded to coat on color after color. The final result wasn't much different than Marco's natural complexion with some red imitation blood dripping from the side of his mouth.

Reaching into the shoe, Marco pulled out the thin paper, about an inch and a half long, folded perfectly about the center. It was his treat from Mrs. Lucas. He read it out loud:

Something special that is both old and new will be your guide to unlock what is true.

Marco's mind shot to Mr. Pikake. He was old, but new to Marco's life. The fortune must be referring to him. Mrs. Lucas never failed to hit the nail on the head.

As he thought, Marco noticed something that rarely happened in his household – silence. His eyes leapt to the clock. The numbers were still obscured by the tape, he pulled his phone from his pocket. Perfect! It was 12:30. Maggie had to be done with her video by now and was probably shut in her room reading the dictionary or some other ridiculous informative.

With anticipation coursing through his veins, he raced down to the office. Beaming when he found the room empty, he slid into the chair and logged onto the game. The sidebar popped up and indicated his friends were not online – Marco would have to take on the mayor himself. He regenerated at the base of the mountain. Alone, Marco had a much harder time battling the stray zombies and avoiding the arrows at the entrance of the cave. After two attempts, he finally was able to squeak through.

Initiating the boss battle, Marco dove in. He ignored the character's face, focusing instead on taking him down. This time, the battle was much easier *without* his team. The room was so small it had forced Oliver, Liam, and Marco into their own corners – unable to move freely without running into each other. Now, Marco had ample room to jump, dive, and attack. After three tries, he leaned back and sunk into the chair having finally defeated the evil mayor.

The man lay in a tangled ball on the floor. His head tilted towards the ground and out of sight. Marco commanded his avatar to climb atop the giant body. The height of the cave allowed the mayor to easily stand. The floor, however, possessed much less room. It distorted the character in unnatural ways so that Marco had to grab and swing to reach the top of the man's knee to find an optimal spot to peer down onto his face. Just then, a textbox took over the screen blocking his view. Marco slammed his hands on the desk in frustration. He read the message.

Mayor: Take the children and your winnings and leave this place! If I see you again, I will not be so kind.

Marco rolled his eyes. "I beat you already dude. Your threats aren't that scary." He muttered to himself before clicking to proceed. Then he saw it. The mayor's face was finally clearly in view. It overtook the screen, staring directly up at him. He leaned in, his mouth hanging open. Maggie burst through the door.

"Fire! Pure fire! Are you watching my video for the week? I bet I go viral. I went in hard on the new bill they are trying to pass. I wouldn't be surprised if Anderson Cooper wants to interview me." She walked around the desk to see what her brother was looking at. "Is that your tutor? Creepy!"

She was right. The mayor bore a striking resemblance to Mr. Pikake. Marco was speechless. Thankfully, Maggie kept talking. "They do that on purpose. They make these characters have faces that look so generic they always remind you of somebody. It makes the game more real. That way, you believe you could actually live in this world."

"Yeah. Yeah." Marco grabbed the crossbow and threw it into his inventory before closing the game. He was surprised how little he cared about it anymore. "That must be it."

"Wanna violently throw snowballs at each other's heads until we can't feel our fingers?" she asked cheerily.

Without responding Marco jumped up and chased her outside. They played until they were too cold to gather anymore snow. Laughing enthusiastically, they ran inside and huddled by the fire to get warm. Their mother brought them hot chocolate and they played Battleship on their iPad until dinner was ready. Peter still hadn't returned, making it the first peaceful meal all week. Marco felt free to scarf down two bowls of spaghetti without his stepfather's evil eyes glaring him down.

That night, as Marco drifted to sleep, his brain was ablaze with thoughts. Coincidence or not, everything was pointing to his tutor. At the start of the weekend, Marco was unsure what to do. The Society of the Abolishment of Numbers, the idea that Mr. Pikake thought Marco could be the next great mind, Peter's threats, and his fracture with Oliver – it was all too much to comprehend. But now, now Marco felt sure that wherever this path may take him, he had already begun his journey down it. There was no turning back. It was like everything in the Universe was pushing him towards one unified point. Deciding to accept his fate, Marco fell fast asleep.

*　*　*

The second week of the year flew by. Things were back to normal, well somewhat. Marco quickly slipped into his traditional spring routine: School. Baseball. Homework. Sleep. School. Baseball. Homework. Sleep.* `

It was early enough in the semester that none of his teachers had assigned any big projects or tests which allowed Marco to easily slide through the school days. After school, Marco and Oliver would always meet at the lockers before heading to practice together. While he knew the rift between them was still there, a pain ached inside when he saw the hallway was empty. He didn't know what he expected…but not seeing his best friend waiting for him was like a punch in the gut.

Oliver was the pitcher for the team, and he was amazing. He was already hitting speeds of 55 miles per hour, and he was just getting warmed up. It was still too cold to train on the field, so the team spent their time running laps in the gym, lifting weights, and perfecting various plays.

Marco's favorite part of practice was when the team completed their base throw drills. Johnathan (the nose-milk boy) was their team's catcher. Oliver would throw the ball fast and hard to him who turned and pitched it to Marco at first base. Marco would pitch it to Henry on second, and he'd pitch it to Colton on third who would ultimately send it hurling back to Oliver. The coach would call out commands to simulate a game. Sometimes, the imaginary runner would try to steal third and Oliver would have to think quick and toss it back to Colton. Or they'd try for a double by having an outfielder pitch it first, followed by Marco quickly tagging the base before sending the ball spiraling towards third or home. They did this over and over again, picking up speed and accuracy.

Always exhausted when he got home, Marco would go through the motions: dinner, homework, bed, before collapsing into a deep sleep. His tiredness was an excellent excuse to keep his head down hoping not to draw any unwanted attention from Peter. The week

---

* In math we call this a permutation of order 4.

was everything Marco needed. There was a normalcy, one he hadn't experienced since before winter break, that he had been craving – but the weekend was fast approaching. While regularity was nice, like a dash of vanilla ice cream, reliable, predictable Marco found himself anxiously awaiting his tutoring session. It was a new and exotic flavor, both alluring and terrifying, never quite sure what he was going to get with each bite.

Mr. Pikake's crazy antics were exactly the change of pace Marco needed after a packed week of school. The math lessons made him feel powerful – how he felt when playing video games. But this was an actual power, not a virtual one. He had the chance to impact the real world, to become a master.

Peter said he was stupid for going to tutoring. On the contrary. He'd be stupid not to take what Mr. Pikake was offering as far as he could. Stacking on the good news, his mother had also instituted a computer schedule because she was, "through listening to the two of you bickering and refusing to share." Marco was awarded mornings to play with Liam and Oliver, and Maggie was stuck with afternoons to record her videos when Marco would be at the local library with his tutor.

On Friday night, as Marco lay on his bed, his mind struggled with his new normal. The PROS and CONS paddles had returned. CONS served up his fights with Peter and Oliver, while PROS returned with his new feelings of control and strength. His focus shifted to Mrs. Lucas' fortune for the year. Old and familiar things seemed to be slipping away, replaced with the new and exciting world his tutor offered him. Plus, he still had Liam, Maggie, and his mother by his side. As the game of ping pong raged back and forth, Marco settled on feeling optimistic about what was to come, allowing his rollercoaster to cruise slowly into the unknown. The dark tunnel that lay ahead lulled him fast asleep.

$$\frac{\left(\dfrac{9}{9}+\sqrt{9}\right)!}{9-\dfrac{9}{9}}^{2}$$

$$\dfrac{3}{\frac{1}{3}}$$

$$\sqrt{81}$$

$$90x - 85$$

$$k + 27z$$

$$\sqrt{522y - 82x}$$

$$7 + 4\tfrac{1}{2}$$

$$20 \bmod 11$$

$$3^{2}$$

$$2^{3} + 1$$

$$\dfrac{10}{1+\frac{1}{9}}$$

9

LIKE TERMS

"THE ESSENCE OF MATH IS NOT TO
MAKE SIMPLE THINGS
COMPLICATED, BUT TO MAKE
COMPLICATED THINGS SIMPLE."

-STAN GUDDER

The public library was only two blocks from Marco's house. As he started his walk, the warm sun shone down while the brisk air was a clear reminder that winter still reigned. He pulled his jacket tightly around himself and tucked his chin down to allow the top of his head to get the brute of the wind.

Marco's thick black hair hung down onto his face just far enough to stay out of his eyes. It made for a natural hat, keeping his head warm so that he rarely needed a beanie. He was relieved when he reached the library, the warm air rushed to escape the room and hit him hard as he opened the doors.

The building was massive. As Marco walked through the metal detectors, to the right he saw what looked like ten tables pushed together to form a U-shape made seamless by long pieces of light wood that adorned the fronts and covered anything hiding beneath. Four workers sat evenly spaced apart like bank tellers beneath a sign that read 'Check Out'. Marco surveyed the space moving counterclockwise, next to the desks were three lines of computers. Only a few were in use by what looked to be college-aged kids. Marco was in awe. He hadn't visited the public library since he was little – he had no memory of just how much lay behind their doors.

Next came the bookcases. Rows and rows going all the way around the room. They reached at least eight feet into the air and repeated all the way to the back walls. Symmetrically, on each side, Marco saw staircases leading to another level that looked like a loft with even more shelves. In the far back corner, he saw a sign that read 'Study Rooms' and could see small chambers each with a table and a whiteboard that hung on the wall.

The middle of the grand entryway was empty. A beautiful mosaic tile pattern created a circular rug and natural light shone down on the design from a series of arched skylights directly above. As Marco's eyes finished their revolution around the room, they landed on the front left corner. A sign read 'Reference' and beneath it was another U-shaped group of desks, smaller than the

$$5(6x + 8x) + 10(2x + 4x)$$

check-out counter. Behind them were short, stocky, bookcases that Marco guessed were only tall enough to reach his chest.

A single librarian sat behind the reference desk, she was chatting and laughing with a figure Marco couldn't quite make out due to the large potted leaves that lay around the room. He stepped onto the mosaic to get a better look. There was Mr. Pikake leaning on the desk, making crazy gestures with his free arm.

Jan. We finally meet. Marco thought. When Mr. Pikake saw him, he wiped a huge smile onto his face and waved his student over.

Meeting him halfway the tutor boomed, "Marco! I'd almost forgotten what you looked like it's been so long!" Although he didn't think he had changed that much in a week, he could agree that it did feel like a long time since they had seen each other. "I reserved us a study room," he motioned to the back corner Marco noticed when he arrived. "How was your week?"

Marco told the professor all about baseball. He shared how he thought they were going to have a really strong team this year and about how sore he was after his first week of practice. He also told Mr. Pikake about his video game and how the mayor looked just like him.

"One of those faces I guess," he chuckled. "Although I do wish they would have made me a nicer character. That mayor seems like a jerk!"

They settled into the third study room on the left. As Marco sat at the desk, Mr. Pikake walked directly to the board and in black marker wrote:

$$40b + 20g.$$

"I believe this is where we left off?" he asked. Marco nodded. It seemed like a lifetime ago they were writing the expression for $40 bats and $20 gloves.

"What I wasn't sure about, was the substitution. Like, if 5 is hiding behind b, does it also have to be hiding behind g?"

$$11(12x - 2x) + 7(3x)$$

"Impressive inquiry to start us off! Let me place the question in your hands. If you purchase 5 bats, must you also purchase 5 gloves?"

"Of course not," Marco was quick to reply.

"Exactly! Almost any Numberfolk could be hiding behind b or behind g. Sure, 5 could hide behind both, or 10 behind one and 15 behind the other. Or 7 behind the other and 12 behind the first. *Almost* anything could be hiding."

"Almost?" Marco inquired.

"Sure! What couldn't be hiding, Marco?"

Marco thought about it. Oliver would have something witty to say like lions, tigers, and bears, but Marco drew a blank. Thinking about Oliver made him nauseous. He shrugged.

"Let me change the question. Can you go to the store, break a bat in half and ask to purchase only part of it?"

The hilarity of the idea cheered him up. If that were true, he could take a handful of Skittles when he didn't have enough to buy the full bag. "I wish," Marco laughed.

Mr. Pikake laughed, too. His laugh was loud and booming. Marco suddenly understood why he got along with Librarian Jan so well. The study rooms were insulated keeping the sound from bouncing all over the open room of the library. The professor could be himself without ridicule here...there was no Head Librarian Sabrina to shush him.

"Sometimes, the context of the problem also helps to restrict the Numberfolk that could be hiding. Makes our job a tad bit easier.°"

° This is called the domain. The domain is all the possible values, er, Numberfolk, that make sense in the problem. In the case of bats, the domain would be all Whole numbers (0, 1, 2, 3, 4, ...) since you cannot purchase partial bats or a negative number of bats, and you certainly couldn't buy π bats.

$$6(9x + 4x) + 6(7x + 2x)$$

"Okay, makes sense. So, anyone could be hiding behind the *b* and the *g*. It could be the same number or different numbers. But since the problem is about bats and gloves, they have to both be wholes, like they can't be one-half or two-thirds, because you couldn't only purchase part of the equipment."

"Superb, son!" He took the eraser that sat on a tray attached to the wall and with a single swipe, the baseball problem was gone. "What I want to talk about today are what we call *like terms*."

Marco had learned about terms last year. It was a mathy way to say whatever came between a plus or minus sign. He quite enjoyed the idea of like terms. He imagined a world filled with expressions, instead of bushes and trees, letters and numbers filled the scene. He saw two *x*'s sitting on a bench, hand-in-hand, falling in love. Afterall they *liked* each other, didn't they?

Wanting to impress his tutor, Marco blurted out, "Terms are the things between plus and minus signs. Like in $5 + 6$, five is a term and six is, too."

"Precisely! Perfectly put." Mr. Pikake drew a remarkably good baseball on the board and then another, he placed the number two in front of each. "If I have two balls and I add another two balls, how many balls do I have?"

The question was so simple it deceived Marco. He was looking for the catch. When he couldn't find it, he went for the obvious answer formed as a question. "You have four balls?"

"Exactly!" He erased one of the balls and replaced it with a bat. "If I have two balls and two bats, do I have four balls? Four bats perhaps?"

Marco laughed. The professor was making a point, but he wasn't quite sure what it was yet. "Of course not. You still have just two balls and two bats."

"I knew you'd pick up on this quickly!" He shot Marco a sideways grin and a wink, "The idea of having bats or balls is the idea of like terms. If there are multiple numbermasks, you can combine them only if they are the same mask." He wrote a new expression on the board.

$$2a + 6a$$

"Tell me son, what can we do with this?"

"I think what you are saying is that since it is the same mask, an a, this is just telling us there are two a's and we are adding six more a's. So just like if there were two balls and we got six more balls, we'd have eight balls all together. Since we have two a's and add in six more a's, altogether we end up with eight a's or $8a$."

"Your aptitude is amazing." He scribbled the answer on the board.

$$2a + 6a = 8a.$$

"Now what about this one?"

$$4a + 9b$$

"Four balls plus nine bats are just four balls and nine bats," Marco replied. He was quickly catching on and he loved the feeling. He was a dark wizard wielding his power.

His tutor beamed with pride, "I suppose I must increase the difficulty for you, young master." He bowed to Marco revealing a gigantic grin when he rose. "This is all there is to like terms. When we have the same numbermask like all x's or all b's or all anything's, we can combine them. But different numbermasks cannot be adjoined."

Marco imagined a Halloween party where everyone was dressed as a werewolf or a vampire. He started sorting the like-masks. There were two werewolves over there, three over here, and five in the corner so that's ten werewolves. A man in a uniform came in and busted the party up saying the establishment had exceeded its maximum werewolf capacity. Marco tried to conceal his grin.

$$8(17x - 3x) + 2(8x + 3x)$$

"One more thing I must warn you of. We spoke a bit about Numberfolk behavior."

"Yeah, the irrational ones," Marco responded under his breath. He thought again about the rabid π in a straitjacket drooling all over the table.

"Numberfolk also function in predictable patterns." He slashed at the board revealing a series of lines and curves. "We shall begin by hunting these," he pointed to the simple line. "Later we will know how to hunt no matter how the Numberfolk functions, and in those cases like terms become more complex." He added to the board:

$$x^2 + x.$$

"Are these terms alike?"

Marco was familiar with the little two and its meaning. This was the E in PEMDAS – exponents. It meant to multiply something by itself and the little number told you how many times. Something like 4^2 meant 4×4 which equals 16. While something like 2^3 meant $2 \times 2 \times 2$ which equals 8. Marco silently scolded himself for his use of the multiplication symbol in his thoughts. He found the baby blanket to be hard to let go of. The question of like terms was more challenging than the appearance of the infant number in the corner. Both terms, x^2 and x had the same numbermask – in that way, they were somewhat similar.

Something was still gnawing at Marco, he had the urge to say 'no' but fought it, "Yes?" he eked out.

"What if I asked you this. If the x were instead bats like we explored before, what would x^2 be then?"

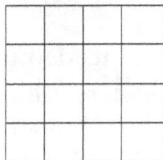

Knowing the square of a number literally meant to make a square (if Marco was asked to find 4^2 he could make a line of four blocks, stack them four tall, and then count the total number of

$9(x + 2x + 3x + 4x + 5x)$

blocks he had) he decided to attempt the same reasoning with the bats.

Crafting a square, the best he came up with was more like a bat-frame. As he studied it in his mind, he felt more and more like this was not at all the same as just a bat, this had a totally different structure to it. You couldn't take the square-bat frame off the wall and whack one out of the park – at least not easily. "I don't know what a square bat would be. Is it bats in the shape of a square or a square-shaped bat? Either way, I don't think bats and square bats are alike."

"Haha, wonderful question. I suppose I thought of it as since 4^2 is four stacks each of length four, that bat^2 would be bat stacks each of length bat." He sketched out his idea on the board. "Nonsensical either way." The professor chuckled before continuing, "We can see that no matter how we imagine the square of bats, it is quite different than a single bat. When a Numberfolk is raised to an exponent it is a legion. It is vastly more powerful than it is alone. The exponent tells us how many are in the army."

Marco imagined going into battle and seeing a single 3 on the field. *Easy*. He thought. Suddenly, a tiny 100 appeared on the 3's shoulder. Marco shifted his view to see it wasn't a single 3 at all! Like a holographic card, when he twisted the army of 3's came into view. They all stood single file, perfectly aligned repeating in unison "I am legion." Marco shuddered.

"When we do not know who is hiding behind a numbermask in an expression like $x^2 + x$, we cannot predict the outcome of their duel – er – their sum."

There it was again, a *duel*. Duels, armies, soldiers, he needed to know more. Marco wished the professor would just tell him already about these Numberfolk duels, but he knew he'd have to be a bit more patient.

$$8(8x + 8x) + 8x$$

Mr. Pikake went on, "There are lots of such tricky situations. These appear to be alike, but do not be fooled! It is only when we have numbermasks that are *exactly* the same that we can combine them. For instance," he scribbled on the board before turning to Marco,

$$5x + 6x = ?$$

"Eleven x," Marco responded proudly. The professor held out his hand indicating for Marco to retrieve the marker. He did, and added to the board before tossing the marker back to the tutor:

$$5x + 6x = 11x.$$

"Wonderful! Let us now try some more advanced substitution. Before, if we had $c + 10$ and 6 is hiding behind c, then…"

"We see 16," Marco finished the thought without prompting.

"Very nice! Things are rarely this clear. All over the world across many professions, humans are battling Numberfolk. And it is unlikely they have poor hiding spots. They are masters of disguise! It may be something more like…"

$$x + y = 6$$
$$y = 2x.$$

Marco stared at the board. This was pushing him to his limits. There were more letters than numbers!

"We call this a system. We will learn much more about systems in time, but this system allows us to increase our substitution skills. It relies on the fact that Numberfolk are all related. Suppose you survey a situation and the pattern you see is $x + y = 6$. I am also analyzing, and I find the pattern $y = 2x$. Together, our information is stronger than it is alone. What is common amongst both our patterns?"

Common. Alike. What's alike here? Marco thought before shouting out. "They both have the same masks, x's and y's."

$$7x + 13(x + 9x)$$

"They do! But even more than that, son. Remember, anything that is equal can be substituted. If $y = 2x$ that means anywhere, anywhere at all when we see a y, that y, is really the same as $2x$'s! We shall hunt two for the price of one!"

"Oh! I see what you're saying. Like if fries and a soda cost $6 and fries are twice as much as a soda, that means three sodas have to be $6." Marco was surprised he had created his own word problem. Never having much lunch money, he was used to this sort of math when weighing his options in the cafeteria.

Mr. Pikake was nodding furiously, his grin like a jack-o-lantern. He tossed the marker back at Marco. The boy approached the board. "Since $y = 2x$, I can replace the y to write…"

$$x + y = 6$$
$$x + 2x = 6.$$

"Yes. Yes. Yes! You replaced the y with a $2x$, since they are equal – the same! Now, what can you do, boy?"

"I have one x plus another two x's. One ball plus two more balls is three balls!" He wrote:

$$3x = 6.$$

"Simply superb!"

"Thanks." Marco smiled softly, "but I don't know what to do from here. I mean, I don't know who's hiding behind x and for sure don't know who's hiding behind y."

"A little more foreshadowing of what is to come!" Mr. Pikake flashed a wicked smile in Marco's direction. "Soon, I will clarify the chase and I will teach you the art of obliteration!" he fired his arm into the air. "But," the professor popped, "I think you would surprise yourself with your innate intuition. You showed me a propensity at our first lesson. Remember the M&M's?"

Marco thought back. He didn't remember everything, but knew they had counted 68 total and Mr. Pikake had turned them into

$$6(6x + 7x + 8x) + 2(x + 2x + 3x)$$

four piles shaped as the letter X. The professor pulled out a small bag of M&M's from his briefcase. "Luckily, I always have a few."

After counting out 6 M&M's from the bag he sat them on a napkin. "Make three piles, each with the same amount of candy," he instructed.

Excitement started to build within Marco, they were back to counting. He was an excellent counter. He didn't even need to touch the chocolate to know the solution. "Two!" He threw his hands up. Skip-counting by two's was easy enough and Marco knew three piles of two would be: 2, 4, 6 M&M's total.

"You see it now?!" Mr. Pikake was dancing around the room. Unfortunately, the small space was just big enough to fit the table in the center and the professor quickly tripped over his own feet. Marco couldn't help being reminded of the mayor from his game. The second consequence of the tiny room was that there was also nowhere to fall. The professor caught himself on the wall before bursting into laughter and dropping into the chair. "*Ah*. You are quite correct. If three x's together make 6, it must be 2 hiding behind x. But," his 'b' popped, "who is hiding behind y?"

The rollercoaster was picking up speed. It wasn't due to fear, it was motivation, determination. Marco was having a wonderful time. He began hunting. "Well, since we know that 2 is hiding behind x *and* that the Numberfolk hiding behind y is two x's, then it must be two 2's! A 4 must be hiding behind y!" He jumped up, tripping over Mr. Pikake's legs as he headed to the front of the room and burst out laughing while adding

$$x = 2$$
$$y = 4$$

to the whiteboard.

Mr. Pikake raised his hand. Despite being seated in a chair, his palm still reached well above Marco's head. The student high-fived the tutor and the two continued to laugh as they reenacted their clumsiness in the small room.

$$7(7x + 9x + 4x) - x$$

"You are doing quite well, Marco." Mr. Pikake said softly after their laughter had died down. "Let me propose a new question."

$$5x + 10xy$$

"Are these like terms?"

Marco wasn't sure. Both terms, the $5x$ and the $10xy$ had an x, so in that way they were alike. The second term, $10xy$, had a second mask. *Did that matter?* Recalling the terms had to be *exactly* alike, he waivered. *If they were alike, what would the result be? Maybe* $15xy$? That didn't make sense to Marco.

"A difficult deduction, *ay?*" the tutor burst into Marco's thoughts. "Let me provide you with a valuable perspective. What does it mean when we see something like xy?"

Marco remembered their lesson on notation, it had been one of his favorites. "It means there are two numbermasks, an x and a y and anyone may be hiding behind these. Also xy means x times y. We don't use a multiplication symbol when there are masks."

"Wonderful work. And ultimately, whomever is hiding behind x and y are just Numberfolk, so xy is simply..." he looked to Marco to continue.

"It's just a number."

"Magnificent! But even more, it is a hiding Numberfolk. Which means it has a mask. Therefore, we need not call the mask xy, we may let $xy = a$. Substitution can simplify the hunt." He added to the board:

Let $xy = a$, then

$$5x + 10xy = 5x + 10a.$$

"Does that help the hunt?"

Now, Marco could easily see the terms were not alike, but he didn't fully understand what had happened. He began to reimagine his werewolf-vampire party. He had already grouped

$$9x(6 + 8) + 9x + 5x$$

the werewolves into one corner and the vampires into another. In the center of the room stood an elaborate hybrid. It had begun its transformation from human to wolf, but it was incomplete. Paws poked out of the guest's suit legs and dog ears adorned its head. Its face was mainly human except for an elongated snout that jutted forward. The guest winked at Marco before slowly brandishing a wicked smile that revealed human teeth except for two gigantic canines that were dripping with blood.

As Marco looked to the left, it was clear the beast did not belong with the werewolves. While it had certain equivalent features, the werewolves were all animal (apart from the formal outfits they wore and the fact they were all standing on their hind legs). Marco then looked to the right, the vampires were mostly human, and the beast certainly did not belong there. He could see how this new creature, while similar to both groups, indeed wore its own unique mask.

"I see it!" Marco exclaimed. "They are only alike if they are *exactly* alike. They can be similar to other masks, and that could be helpful to know, but they are *not* the same."

"And now, you are ready to begin formally hunting." Mr. Pikake raised both arms symbolically lifting Marco to the next level of training. "You have become a master of the letters, a master of the numbermasks, Marco. You should be proud."

And he was. Before working with his tutor, the mixture of letters and numbers were terrifying to Marco. Now, he saw how ridiculous it was to fear a letter. Yet an unmistakable feeling, a sharp and steep drop on his internal rollercoaster, still reared its ugly head anytime these hybrids appeared.

"Next time, son, I will teach you of the duels. We will begin our formal training on the hunt."

Marco was instantly filled with excitement and anticipation. He had been patiently waiting, forming his preliminary skills, and now it had paid off. It was finally time.

$$2x + 7x + 8(2x + 7x) + 6(x + 9x)$$

"First, I want to share with you a difficult design. When we hunt, it is our job to identify the Numberfolk who is trying to deceive us."

"Yeah!" Marco was so excited he couldn't hold back, "Like $3x = 6$, the two was hiding."

"Yes, son... but this." He paused trying to find the perfect words, "In terms of conceivable covers, this is a more basic veil. I must show you how complex their camouflage can become. How skilled the Numberfolk can be with their costumes."

Fear crept into Marco's throat. His imagination took over. He tried to stop it, but his mind ignored him. He saw Maggie, jumping up and down on his bed, her normal giggles bouncing along with her ringlets. Slowly, her smile twisted into a crude grin. Marco recognized it, it was the crazy 5 who had been attached at the hip to the fierce 1. He stood in shock as his sister transformed into 15, laughing demonically. He squeezed his eyes shut trying to push the thought from his head.

"When you know what they are capable of – their many forms – you will be better able to conquer them." The professor's voice helped pull Marco back to reality.

"Okay," Marco said wearily. "Show me."

Mr. Pikake jumped to the board and scrubbed it with his forearm. Marco watched as his tutor made the same motion over and over again before stepping to the side to reveal his work[*]:

---

[*] In case you need a refresher, a decimal is just another way to express a fraction. The fraction $\frac{1}{2}$ is equal to the decimal 0.5. Some decimals like 0.5 are terminating, meaning they end. Since we use a place-ten system, anything that is divided by a multiple of 2 or 5 will terminate. Like $\frac{1}{8}$. Since 8 is a multiple of two, it will terminate as 0.125. Other decimals are repeating, like here, they go on and on forever. Irrationals are different kinds of decimals, ones that can't be expressed as fractions, but you'll have to wait to learn that crazy story.

$$11(9x + 3x) + 8x + 2x$$

$$0.999999\overline{9} = 1.$$

Marco's head actually exploded. The wave of questions and uncertainty that washed in like a great flood was too much for the boy to wade through. As he struggled to tread water, he grabbed onto the first question he could as a life preserver and blurted out, "What is the line over the last nine?" While not his most pressing question, it was all he could manage at the moment.

"Excellent inception! In a decimal, a line atop tells us it continues forever. You see Numberfolk have the ability to be never-ending, immortal."

Marco would not soon forget the irrational π thrashing and clawing away. He shook his head to keep its image from forming in his mind.

"We humans do not possess such a skill, we cannot even represent such Numberfolk for our studies. It would require us continuing to scribble nines *forever*. Inconceivable! To not waste our time on what cannot be done, we instead use this simple mark." The professor provided another example on the whiteboard:

$$0.\overline{34} = 0.343434343434 \ldots.$$

"When we add the line, in this case over the three and four, it means this Numberfolk repeats itself as three-four, three-four, three-four, for all of eternity and then some. It is what we call *infinite*."

Thankfully, Marco was familiar with infinity… as much as a seventh grader could be anyway. Infinity to him was a Hot Wheel's track. Two loops that once he set the car loose on, would keep going and going and going.

The flash flood had subsided and while the waters were still deep, they were calmer now, allowing Marco to begin

$$9(9x + 3x + 7x - 4x) + 9x - x$$

deconstructing the situation with more clarity. "Okay, nines forever. Got it. But that doesn't mean it's one." Marco surprised himself with the firmness of his tone. Afterall, he wasn't a math kid and Mr. Pikake was a grand master of Numberfolk. But he couldn't help it. Something inside of Marco felt sure that a bunch of nines were definitely different than a single one.

"I shall take a crack at convincing you. May I ask – how much is three-thirds?"

Not seeing the relevance, Marco decided to go along with this new train of thought, "Well, that *is* one. If I cut a pie into three pieces and then eat all three, I ate the whole thing – one whole pie."

"Very true. And we can write one-third as the decimal," finishing his sentence on the board he scribbled[*]:

$$0.333333\overline{3}.$$

"If,"

$$\frac{1}{3} = 0.333333\overline{3}$$

"and,"

$$\frac{1}{3} + \frac{1}{3} + \frac{1}{3} = \frac{3}{3} = 1$$

"can we not say,"

$$0.333333\overline{3} + 0.333333\overline{3} + 0.333333\overline{3} = 0.999999\overline{9} = 1?$$

Marco saw what his tutor was doing, he was using the substitution he had taught him. As he studied the writing, he felt like reality was tearing apart. It didn't make sense and yet it also

---

[*] If you type in $1 \div 3$ into some calculators, they might lie and tell you something like 0.333334. They do this because they are rounding. Just like humans, they can't go on forever so they stop trying at some point. The good calculators will give you 0.333333 as to not confuse you.

$$4(x + 2x)(3(2x + 2x))$$

made perfect sense. Marco accepted that the decimal form of one-third was $0.333333\overline{3}$. He also knew that if you added together three of those, you would get $0.999999\overline{9}$. On the other hand, three one-thirds were certainly one whole. As much as he tried, his brain would not accept that an infinite number of nines after a decimal point was equal to one. He felt like this was some elaborate magical illusion work. Mr. Pikake had cut a woman in half in front of him yet she lay there smiling and talking like nothing had happened. Marco found himself searching for the hidden mirror that made the trick possible.

"I can see I have not yet convinced you," the professor read Marco's thoughts. "I am showing you this for three reasons." The first of his bony fingers shot up, "One – it is a very good instance of a Numberfolk disguise, perhaps one of the best! To be a spectacular hunter, you must understand the many ways they will attempt to thwart us!" The next skeletal digit sprung to life. "Two – this provides a fine introduction to the basics of hunting. And three," his thumb wrenched to the side making his hand appear alien like the statements on the board, "this forces us to widen our minds, to be Numberfolk Jedi who are able to wield the invisible powers of the Universe."

Marco took a deep breath. He desperately wanted to be a Jedi, a master. It wasn't about math, or school, or grades. He could feel the great power the professor was offering him, and Marco craved it. Looking Mr. Pikake dead in the eyes he firmly replied, "Teach me."

A wicked grin swept across the tutor's face as he continued.

"Let us take on a different perspective. Say this Numberfolk is hiding behind the numbermask a."

$$a = 0.999999\overline{9}.$$

"What would ten of the a's appear as?"

$$15(3x + 3x + 3x) + 5(x + x)$$

Thankful for a question he knew the answer to, Marco retorted, "Ten times a decimal means we just move the decimal place to the right one space." He liked using money to think about it. If he had a dime, $0.10, and then collected nine more dimes to have ten in total, he would have $10 \times 0.10 = 1.00$, a dollar. He admonished himself for using the multiplication sign again in his thoughts.

"Perfect! Therefore,"

$$10a = 9.99999\overline{9}$$

"agree?"

Marco nodded, this made enough sense.

Excellent! Now, let us subtract! On the left we have $10a - a$, how many numbermasks does that leave us with?"

Ten balls, I give one ball away, I have nine balls left. Marco thought before blurting out, "Nine, we would have nine a's."

"That we would! Now, for the right. The ten a's were $9.99999\overline{9}$ and the single a was $0.999999\overline{9}$, let us subtract these as well."

Wanting to be certain, Marco dug out a piece of paper and a pencil from his bag and wrote down the problem:

$$
\begin{array}{r}
9.9999999\ \dots \\
-0.9999999\ \dots \\
\hline
9.0000000\ \dots
\end{array}
$$

"Okay, we have just nine. All the decimal places become zero, so you get nine point zero zero zero forever."

"Precisely!" The professor scribbled on the board:

$$
\begin{array}{r}
10a = 9.9999999\ \dots \\
-(a = 0.9999999\ \dots) \\
\hline
9a = 9.0000000\ \dots
\end{array}
$$

"If nine of some numbermask is equal to nine, who is hiding behind the mask, son?"

$$21x - x + 21(2x + 2x + 2x)$$

Marco didn't need to understand duels or whatever the obliteration his tutor had mentioned today was. He knew that the only thing he could have nine of and end up with nine was one. That didn't require any special counting. He had nine M&M's, he made nine piles. Of course only a single M&M would be in each mound, 'fracturing' as Mr. Pikake called it.

"One. It has to be one hiding behind the a," Marco replied feeling defeated.

"Very good, boy," the professor said kindly. "And if 1 is hiding behind a and $a = 0.9999999\overline{9}$, then," he wrote the bewildering statement once more on the board:

$$0.9999999\overline{9} = 1.$$

He stopped treading water and allowed the waves to pull him under. As he sank deeper into the ocean of questions, he surrendered to the uncertainty. Marco was *sure* the two numbers were not equal. Yet, Mr. Pikake was able to show in multiple ways, ways that made sense, that they were in fact the same.

He felt something on his ankle. Bending over to swat it away, he saw it was a chain. Attached to the chain was an enormous anchor that had clear letters stamped onto its side: NOT A MATH KID. The weight was pulling him down. Trying to fight it he thrashed, tugging at the water, and manically kicking his free foot. The force was simply too powerful to overcome. Just when he thought he was toast, a hand clamped onto his wrist. He looked up to see Mr. Pikake lifting him from the deep.

"Do not fret, boy." Emerging from his trance, Marco saw the professor still firmly holding onto his arm. "I show you this to enlighten you. To help you realize just how slippery the

$$3(7x + 7x + 7x + 7x) + 7(3x + 3x + 3x)$$

Numberfolk can be. It will be some time before you encounter such clever charades. I want you to keep it in your mind that they are trying to trick you. It will help you on your hunt." He released the boy's arm and bent down to retrieve something from his briefcase. "I know, as much as I know anything in this world, that you are capable. You will succeed, Marco."

He revealed the item to his student. Like the book the professor had shown Marco before, he now held a miniature version bound in soft leather with thick jagged pages poking out of the binding. He continued, "This is my hunting journal. In it, I have recorded all my battles. Not only to reminisce on the victories, but to learn from my failures."

Marco was surprised. He had assumed a grand master like Mr. Pikake wouldn't have ever failed. He wasn't like Marco. The professor was like Oliver and Maggie – math kids. The idea that the talented man had been defeated before was comforting, it gave Marco hope.

He reached back down into his bag again and yanked out a second book – the journal's twin. "I want to present you with your own hunting journal. It is yours to fill with your own adventures."

Allowing the tutor to place the book in his hands, Marco took the long rope attached to the binding and untwirled it to look inside. Each blank page was constructed of a heavy, bumpy stationary that begged for ink. Marco saw the book filling with his own triumphs, and his own disasters, and he was overcome with anticipation.

Opening his own journal, Mr. Pikake slid the book across the table and nodded to Marco to look inside. "I was only slightly younger than you when my father presented me with my first journal. I shall share with you what he shared with me. He instructed me to begin by explaining a numbermask, a variable, a letter. Our power lies greatly in understanding the enemy. So, I say to you also begin with everything you know, everything we

$$3(10x + 11x + 12x + 13x) + 10x$$

have learned about the masks. When in doubt, you will always have this to refer to and help to guide your way."

"Th-th-thank you," was all Marco could say. He was transported back to November when Liam celebrated his 13th birthday with a bar mitzvah. Marco had watched the beautiful ceremony and admitted feeling jealous when he saw how his friend's father doted over his son, passing down a family heirloom. Marco knew he would never have such a moment. Now, receiving this gift from Mr. Pikake was as close as he would ever get.

His eyes swelled with tears as he pushed out another, "Thank you." He knew this path would not be easy, but Marco was filled with more determination than his child-like body could hold. He *would* make his tutor proud.

Jan waived furiously at Mr. Pikake like a teenager at a pop concert as they left the library. When they exited the building, the heated air pushed them from behind forcing them into the frigid temperatures outside. The sun was starting to set and there was no warmth to be found.

"Until tomorrow," Mr. Pikake said pulling his suit jacket tight around his waist. Marco waived and began the jog home. Still running, he burst through the front door and up the stairs. He couldn't wait to record his first entry in his journal.

He grabbed a dark ink pen from his desk but hesitated. There was nothing worse than opening a new toy and instantly defacing it. It still made Marco mad when he thought about his new LEGO set. It came with shiny stickers to affix to certain parts. "Let your sister help," his mother insisted. In minutes, baby Maggie shoved a sticker on, wrinkled and crooked. Marco carefully tried to surgically remove the decal to fix her mistake, but the adhesive was too strong. Lines of white paper remained on the block, ruining it, forcing it to never reach its full potential.

Not wanting the journal to have the same fate, he thought over everything he had learned from Mr. Pikake before carefully beginning to write.

$$10(10x) + 7(7x)$$

Part 1: Letters

Letters are hiding places for numbers. They are masks and it is the duty of the hunter to uncover who is cloaked beneath.

When a letter and a number are together, it tells how many masks there are. A 4x means there are four x masks. This can be x+x+x+x or 4.

Marco quickly pulled the pen from the paper. Realizing he was about to write the multiplication sign he sighed, thankful he stopped in time. He read over the last line before continuing, 'This can be $x + x + x + x$ or $4'$

times x.

Terms can be combined when their masks are exactly the same. If there is 4x+2x, this means there are four masks and two more of the same mask are added meaning there are six identical masks together, 6x. If the masks are not the same, they cannot be combined. I cannot simplify 4x+2xy.

Looking over the page, Marco was happy with his work. But something was missing. He needed to create his own world, a world where all of this was in his terms, in ways that he understood. Knowing just what to add, he continued.

If the x are werewolves and the y vampires, the xy is a hybrid beast. It belongs not with the x or with the y.

Next, he drew in sketches of the creatures to complete his visualization.

He finished with a note on patterns.

$$5(9x + 10x + 11x)$$

uncovering patterns is an important part of hunting. If the pattern is 20 - x, I can test what I would see for different hiders. If 1 is hiding, I would see 19. If 2 is hiding, I would see 18. Now, I know the pattern. The bigger the one hiding, the smaller number is seen.

Lastly, **substitution**

Marco retraced the letters again and again so they would jump off the page.

is a powerful tool. Anything equal can be substituted. If y=2x, anywhere there is a y, I can replace it with a 2x since they are the same - equal.

12 IL   20-x   ID 1L

Completing the page by doodling a number factory to display a pattern transformation, Marco set down his pen. He scanned the page again and again. Happy with his work, there was still something missing. Then it came to him. In big letters at the very bottom of the page he added:

YOU ARE NOW READY FOR THE HUNT.

$$14(9x - 2x) + 5(9x + x) + 3x$$

DECODING UNIT 3

$$4(3x + 6x + 12x) + 4(9x + 5x + 3x)$$

It's been so long since we have encountered a blank page... Unfortunately, that 'not truly empty' page on the left is looking at me funny.

Here's a solution! I'll provide a riddle. You can use page 152 to work it out. This way, we can ensure it won't stay empty.

Numberfolk have many human traits. They can be magical, happy, impolite, admirable, powerful, lucky, frugal, or upside-down, to name a few. The Naturals 128, 175, 258, and 261 each have two of these traits and no two of these Numberfolk share a trait. Can you determine what belongs to each one?

1.   The admirable Numberfolk is even.
2.   The lucky Numberfolk's digits sum to Nava's Spogs.
3.   The powerful Numberfolk is not the smallest.
4.   The Numberfolk who is magical is also powerful.
5.   128, the upside-down, and happy are all different Numberfolk.
6.   Upside-down numbers are those where their outer digits sum to their middle digit doubled.
7.   The frugal Numberfolk is so rude!

Each of these traits is a very real mathematical concept. Knowing this may make the next chapter seem not so far-fetched after all...

$$12x + 12(12x) - 3x$$

10

A BRIEF HISTORY OF NUMBERVILLE

"YOU DON'T NEED TO BE A MATHEMATICIAN TO HAVE A FEEL FOR NUMBERS."

-JOHN NASH

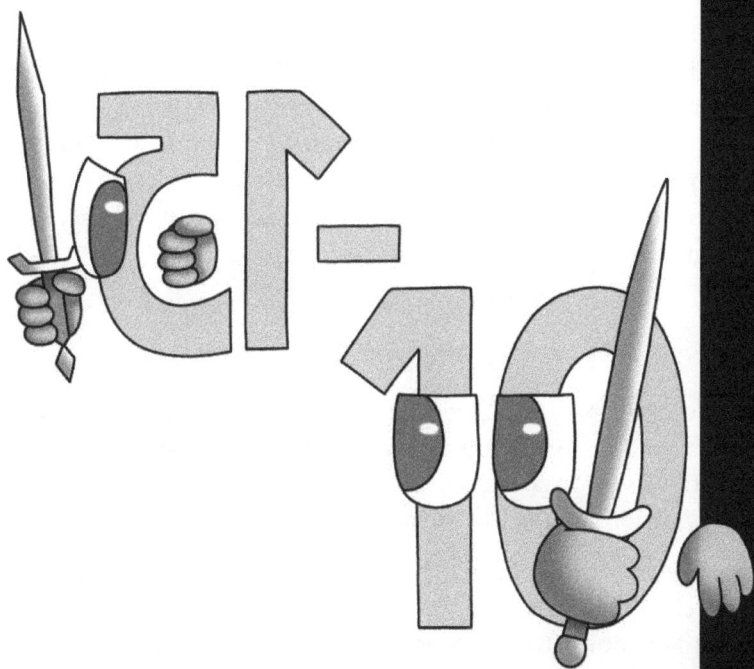

Natural was a bustling village within the land of Numberville. A beautiful place, it contained lush green hills that stretched much further than the eyes could see. After its construction, the ten charmed figures left to allow their creations to learn and grow without interference.

What we consider counting numbers $(1, 2, 3, 4, 5, 6, \dots)$ populated the village. While Zil had worked with his siblings to form an infinite number of chains $(10, 20, 30, \dots)$, his image – zero – hadn't been seen since they arrived in this new world.

While Natural had all the charms of a small town, a quaint central square, and modest homes for each resident, it was in fact a metropolis. If one wished to catalogue every resident, they need only to continuously count, adding one each time. Unfortunately, this would take well past forever. Deciding they were countable, yet infinite, they erected a sign[*] on the border that read:

VILLAGE OF NATURAL

POPULATION: $\aleph_0$

The residents of Natural lived in harmony for many years. Yet as time passed by, the Numberfolk became restless and bored. Bickering in the streets was soon a daily occurrence – they needed a place to relieve their tensions and settle their disputes. Thus, dueling began.

At first, when two residents had a dispute, Numberfolk would watch on as they yelled and screamed until someone eventually gave up and went home. This never truly settled anything, and they were generally back the next day arguing again. In an attempt to resolve these disturbances, the residents came together to brainstorm a solution. "The larger should be declared the victor," a ten thousand four hundred and twenty-two suggested.

---

[*] They called the number of residents in Natural 'aleph naught'. It is a way to represent infinity. While it is never-ending, it is also countable and represented by the symbol $\aleph_0$.

A brave six refuted the proposition, "Simply ridiculous! Your size does not determine your accuracy!"

After days and days of arguments, dueling was proposed. With no better options, the Numberfolk begrudgingly agreed and set forth to establish the ground rules. When two residents had a disagreement, they would meet in the town square. Each would walk exactly ten paces – one to the east and one to the west – before turning to face each other. At the sound of a neutral party's call (everyone unanimously agreed this would always be a one, for One had something in common with all Numberfolk), the two residents would run at each other headfirst. The Numberfolk left standing would be declared the winner.

Mind you, this was as ridiculous as things come. No one had any idea what would happen when two Numberfolk collided in this way – it had never been done before. Grumblings about fairness also rang through the village. When a twenty-four dueled a five, would it really be a just fight? As things had become so strained, and since with a duel the five would at least have a chance unlike some of the other suggestions, everyone reluctantly agreed to the rules.

The first to ever duel was a fourteen and a twelve. They lived in the same neighborhood within Natural with a thirteen residing between the two. Thirteen was a horrible neighbor, she never budged on anything and seemed to lack the ability to compromise. The argument between Twelve and Fourteen all started when they were discussing their common neighbor.

"She is just incorrigible!" Twelve started. "It's all because of her makeup. You know, she can't be split into equal parts at all. My flexibility is my greatest trait. I am three four's, or four three's. I can be two six's or six two's. Your ability to mold to fit your surroundings is everything these days."

Fourteen was horribly offended by Twelve's comments, after all, he could only manipulate himself into two seven's or seven two's. "That isn't true at all!" his voice was already raised. "I don't have many forms, but I am still an excellent neighbor!"

Twelve fired off, "You trying to say you're better than me? Impossible!" And solely on the basis of who was the perfect neighbor, the inaugural duel was declared.

Numberfolk came from far and wide to witness the event. The town square was filled to its capacity with residents. Things were so tight, a one and a two seated next to each other were almost pulled into the duel mistaken for a twelve.

Twelve and Fourteen made their way to the center of the square. The crowd went silent, every resident filled with anticipation. They each took their ten paces, Twelve to the east and Fourteen to the west. They turned to face each other. One was the neutral party that yelled, "BEGIN!"

Before anyone could blink, the two were sprinting directly towards each other. As they crashed, everything appeared to be going in slow motion. Both Twelve and Fourteen were fractured, ripped at the seams. Twelve went up and split into a ten and two ones, Fourteen split into a ten and four ones, their pieces floated through the air. Time sped back up and like a black hole collapsing in on itself, the pieces were sucked together in a plume of smoke.

No one said a thing. The residents of Natural all stood in awe at what they had just witnessed. Somewhere, someone began to cry. Other residents looked at each other, perplexed. What happened? As if an unseen force pressed unmute, everyone began yelling simultaneously.

"What have we done?!"

"How could we have let this happen?!"

Others were pointing fingers, "It's all your fault, you thought the duels would be a good idea."

And still others claimed to have somehow foreseen the entire event, "I knew this would happen. I mean, what did you expect?"

Suddenly, a voice rang out above all others. It was One. She was pointing to the middle of the square, "Look! Look!" she screamed. As the dust slowly cleared, residents pushed and squinted to see what remained of the dueling Numberfolk. At the very spot where Twelve and Fourteen had collided stood a Twenty-six. The crowd erupted into cheers. From that day on, Numberfolk loved to duel, be it to settle a disagreement, for entertainment, or simply to explore the result. Not a day went by when two Numberfolk, if not more, could be seen dueling in the square*.

In a twist of irony, the result of Numberfolk duels were always a new, single, united Numberfolk that was greater together than they were apart.

* * *

The next major event to occur in Numberville was the Integer Wars. The counting numbers knew very little of the world outside Natural. They knew Natural was in the county of Whole which lay in the state of Integer. Many believed they also lived in the province of Real, but having never met anything that wasn't real, this was more myth than fact.

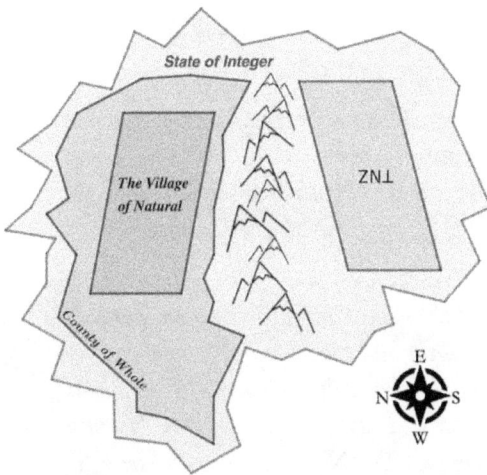

Whole, outside of the village of Natural, was a dismal place, very little grew there. It was nothing more than a vast desert with a single inhabitant: Zero. While all the Naturals technically lived in the county of Whole, Zero was the only Numberfolk who lived in the county of Whole, but *not* in the village

---

* Today, most schools refer to Numberfolk duels as *addition*. A politically correct way to cover-up the torrid history. The word addition comes from the Latin meaning to join or unite – a much more positive spin on Numberfolk violently hurling themselves at each other.

of Natural*. He enjoyed the solidarity and for a long time, he didn't have to worry about silly neighbors or duels.

Zero spent his time travelling through Whole. He walked along the outskirts of Natural and decided to head south to see what secrets lay there. Intentionally saving this trek for last, all Zero could see to the south was a treacherous mountain. As he stood at the base and tilted upwards struggling to see the peak, it seemed like an insurmountable task. His curiosity could not be deterred. Zero began his assent.

When Zero reached the summit, he looked down to the north and saw Natural. It seemed almost small from where he stood. He then turned to the south and saw Natural again. It was as if he was looking into the Mirror of Wonders herself. He couldn't believe his eyes. He studied the village to the south. Although remarkably similar to its twin to the north, it was not Natural – it was in some way different. Unknown to Zero, he was gazing down on the creation of the Great Scale. The longer Zero watched, the more he realized it was the exact *opposite* of Natural.

Zero camped atop the mountain for many days and recorded everything he witnessed.

> *When I set forth on this journey, I had no expectations. I travelled the land of Whole and saw only the village of Natural, its inhabitants, and myself. If anything, I expected more of nothing when I reached the peak of the great mountain. What I found was simply fascinating.*
>
> *I have named the village identical but also quite different from Natural that lies to the south of the mountain, The Negative Zone. It didn't seem fit to call it Un-Natural, and the town brings about in me feelings of dread, of distrust. For the time being, I refer to the residents of The Negative Zone as 'evil twins.' I look to my right into Natural and see 1, I look to my left down into The Negative Zone and see −1.*

---

* Our modern geography intentionally mimics that of Numberville. A resident of Chicago may live in the city meaning they are also a resident of Cook County and others may live in Cook County but outside the city of Chicago.

$$300 + (-140)$$

Like looking into a pond, 1 and −1 are brothers, nearly identical, but also very different.

The residents of The Negative Zone appear wicked. They are not kind to their neighbors and even the slightest offense leads to a duel. They are pugnacious, always ready for a fight, and for the short time I sat and observed, plumes of dust — the ruminates of a duel, burst as far as I could see from all edges of the town. What is even more interesting is that in Natural, while disagreements and duels occur, the result is always a resolution, something larger and stronger than what went in. But in The Negative Zone, the duels are more malicious. For if two members of this world duel, the result is always smaller. They diminish each other both tugging down closer to the hellish abyss.

I feel as though if I could fold the world in half about this peak, each side would align perfectly, a horribly evil creature with its wonderfully pleasant twin. The result I think would be spectacular, there is a beautiful symmetry between the two — for that I am certain.

The combination of Whole including Natural and The Negative Zone (TNZ) made up the state of Integer. Once word got around of the 'other side of Integer' in almost no time at all, a civil war was declared. Naturals were terrified by Zero's account of these 'evil' twins. Of course, in TNZ, they saw the Naturals as imposters having stolen what was rightfully theirs.

At this point in history, Numberfolk civilization was fairly young. They had barely figured out how to settle disputes between their friends, much less with an unknown enemy. They decided it seemed fit for each resident to battle its evil twin. (To the Naturals, the Negatives were the *evil* twins and to the Negatives the Naturals were, well, naturally evil). This became known as the Battle of Cancelation.

For the first duel in the Battle of Cancelation, Natural sent 1 and TNZ sent −1. Both sides waited anxiously to see the result. Much time passed and to the dismay of both lands, no one returned home. Trying again, Natural next sent 2 and TNZ −2, repeating their past yet expecting a different outcome. Every Numberfolk was disappointed as once again, no one returned. This continued for some time: Natural and TNZ repeatedly sent their twins to duel, until Zero finally intervened.

He went to each border and explained what the battles were producing. While there existed many of each Numberfolk (multiple ones, twos, threes, etc.) there was only one Zero. The single image of Zil himself, Zero loved the solitude of Whole. The Battle of Cancelation changed everything. For every duel that transpired resulted in a new zero. When 1 battled -1, a zero was the result. When 2 battled -2, another zero. Now Whole was filled with zeros everywhere! Both sides came to understand why no Numberfolk had ever returned home. Zero was not a resident of Natural nor of TNZ, a zero could only reside in Whole and thus the Battles of Cancellation created a purgatory, never allowing the soldiers to leave.

Neither side could justify sending twins to duel again. They each devised their own new strategy. Legend has it that the Naturals first sent 10. One of the most popular residents, ten's were viewed as pillars of strength in the community. This was a product of their Spog ancestry, the fact that there were ten original charmed ones – ten digits – made ten's heralded above the other Numberfolk.

TNZ picked -15 and hoped for the best. The Naturals were shocked when no one returned after the duel, while TNZ was pleasantly surprised when they saw -5 descending the mountain. This was a turning point in the war as the Numberfolk began to understand their properties. The battles continued. Natural sent 32, TNZ sent -12, 20 returned to Natural. Natural sent 57, TNZ sent -100, -43 returned to TNZ. Seeing only tragedy in the wake of the battles, a cease fire was declared and all of Integer agreed to adopt a more diplomatic resolution.

Enter the Thinkers. The Thinkers were constructed as an elite group of Numberfolk composed of residents of Integer. An election was held that awarded three seats to Naturals and three seats to Negatives from TNZ. The final, seventh, seat was reserved for a zero. The goal of the Thinkers was to study Numberfolk phenomena and publish their results. Their first agenda item was to explain the results of the Integer Wars.

After much study and deliberation, the Thinkers present the Integer Papers. The purpose of this document is to bring to light the underpinnings of dueling between Natural and TNZ as occurred in the first war.

Our process began by taking a census of the residents of Integer, organizing them, and then analyzing the previous dueling results.

The residents of Natural are composed of positive particles matching their value. Thus, any duel between two Naturals is simply the union, or combination, of these positive particles. When a 5 duels a 3, the outcome is an 8.

Functioning in an opposite nature, the residents of TNZ are each composed of negative particles. Hence, any duel between two Negatives is the union of these negative particles. When a −4 duels a −6, the outcome is a −10.

This understanding accounts for Zero's initial observations as well. Duels between Naturals result in more positive particles – a larger value, while duels between Negatives result in more negative particles – a smaller value. The symmetry of our two lands is apparent. The twins of −4 and −6 are 4 and 6 respectively. If they were to duel in Natural, the result would be none other than 10, the twin of −10 and the result of the symmetrical duel of −4 and −6 in TNZ.

We then turned our focus to the duels of the war. How can we explain the outcomes of the battles between Natural and TNZ? The Battle of Cancelation showed us that when any Numberfolk duels their twin, they are vanquished – reduced to nothing. We found a similar outcome with the wars. Take the skirmish between 10 and −15 as an example. A −15 can be thought of as the fusion of ten and five negative particles. The duel then consists of 10's positive particles battling −15's ten negative particles and the tragic result was none other than cancellation, they were vanquished. As −15 retained five negative particles that were not used in the duel, the ultimate outcome of combat was, −5.

Our findings clearly spell out the need to cease any military answers to our disagreements. The victor is not the better, the stronger, or the wiser of the two sides, it is but chance. It's the guess of the generals who to send into combat and for any resident of either land, there will always be someone bigger. It is the opinion of your elected Thinkers that the wars be ended promptly, and diplomatic methods are adopted to resolve our conflicts.

The residents of Integer appeared to take the advice of the Thinkers, and a time of relative peace – a Pax Numera – was enjoyed throughout the land. Behind the scenes however, tensions continued to build. Many Naturals saw themselves as the rightful residents of Integer. Having been created by the charmed ones themselves, they saw TNZ as a horrible mistake. The Negatives were errors, unintended consequences, and as such, should be eliminated.

Out of this, guerrilla warfare was born. As the strain between the lands increased, the battles resurrected with a new strategy at the forefront. It was customary, as with all duels, that the combatants would take ten paces in opposite directions (each towards their own motherland) before turning and racing to collide.

One of the great commanders of Natural, 467, had a new idea. She sent out a 41. Before its journey, she pulled the warrior aside and provided their orders. Knowing that if TNZ sent out the twin of anyone larger than 41 TNZ would be the victors, she instructed her soldier to do the following. "When you arrive, first determine who you are facing. If their twin is larger than you such as $-42, -43, -44$ and so on, you shall know before the battle even begins that you will lose. Your only chance is to do exactly as I say."

She then instructed 41 to turn as if to take their ten paces but then immediately revolve and attack from behind! This was a clear violation of the conventions, but 467 didn't care. If they were going to be vanquished anyway, this seemed like their only hope at survival.

As 41 approached the battlefield, they quivered when they saw -600 standing before them. Doing as their commander had asked, they took only a single pace before running to attack. Anxious to know the result, 467 waited at the border of Natural, but no one returned.

When she heard the loud cheers from TNZ she knew her plan had failed.

Negatives stood in shock when their warrior came down the mountain. In every previous battle the soldier who returned, if anyone returned at all, was always a larger value, they were less negative having lost some of their particles in the fight. But this time, -600 went into battle and -641 returned. How was this possible? As -641 described the battle, TNZ was more than intrigued to test out this new strategy.

The next duel was between 320 and -61. The -61 was instructed to backstab and they did. No one returned to TNZ but, surprisingly, 381 returned to Natural. Things were getting messy, countless Numberfolk were vanquished as each side pushed the limit of this new strategy. It was the skirmish between -23 and 46 that finally put an end to things.

Natural had no reason to use the backstab defense, this was already 46's battle to win. Being new to fighting, scared, and unsure, when the duel began both Numberfolk turned and attacked, neither taking their paces to engage in formal combat. As -23 returned home to TNZ silence rang out. Never before had a resident returned without any change! For war always has an impact on the solider. Expecting to be rejoiced and given a warm welcome, -23 walked the streets of TNZ. No one cheered. Residents looked away, they hid, or they simply stared in awe. What was this witchcraft -23 had engaged in?

A boycott of the war ensued. Residents refused to fight and demanded answers to better understand this new strategy. The Thinkers went to work.

> The new battle strategy in which both Natural and TNZ has willingly broken the rules of engagement has been quite unsettling. The results have been strange, unpredictable, and honestly terrifying. We have studied the new methods and come to a conclusion.
>
> 'Backstabbing', as it has come to be called, produced some unintended consequences. The first documented use of this technique was the confrontation between a 41 and a -600. Rather than duel following

$$-(-56 + (-109))$$

the standard ordinance, in this case, **41** used the backstab technique and the result was surprisingly **−641**. In order to explain the results, we have composed two diagrams. The first, demonstrates a traditional duel, notice each resident is willingly engaged in battle.

−600 ⚔ 41

Figure 1: A traditional duel between -600 and 41.

As we know, **41** positive particles vanquish **41** negative particles, and the outcome of the battle would be the **−559** particles that remain.

Now consider the new backstabbing strategy. Here, only the **41**, the backstabber, is engaged in battle. We have found backstabbing has the effect of removing particles. Therefore, in this case, **41** particles were removed from **−600** resulting in the outcome of **−641** particles.

−600 ⚔ 41

Figure 2: The backstabbing defense. Here -600 is engaged in normal battle while 41 has backstabbed his opponent.

The Thinkers have found this entire strategy to be disturbing. It's as if we are willingly giving ourselves away in senseless battles! For history, academic purposes, if only to deter this practice from ever happening again, we leave you with this. The backstabbing defense is equivalent to dueling with one's twin. When **41** used the backstab technique, it was as if **−600** dueled with **41**'s twin, **−41**. The result was, of course, the smaller **−641**. In a duel with their own, this method would have the intended consequence. That is, should **41** backstab **600**, both Naturals, the result would tear down the **600**, removing **41** particles from him. But as TNZ enjoys being less while Natural relishes in being more, this activity is nothing more than a kamikaze sacrifice.

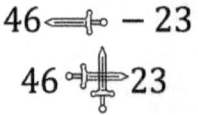

46 ⚔ − 23

46 ⚔ 23

Figure 3: When -23 backstabs 46, the result is the same as a traditional duel with 46 and 23.

Many questions have been floated regarding the battle between **−23** and **46**, in which both sides used this deplorable attack. When **−23** used the backstab strategy, the result was a duel with **46** and **−23**'s twin, **23**. This would conclude with **69**. However, as **46** used the same technique, this became a duel between 23 and the twin of 46 which is of course −46. Therefore, the battle became one with 23 and −46 which resulted in −23, there was no witchcraft or other methods involved. [*]

Figure 4: When both 46 and -23 backstab, this is akin to a duel between their evil twins.

The use of backstabbing has become a dilemma of war. Suppose we place two residents into the field, we shall call them resident N and resident G for anonymity. Resident N is a Natural and resident G a Negative. There are four possible outcomes.

- Both N and G decide to abide by the rules of warfare and engage in a traditional duel. In this first case, the traditional duel, whomever possessed more particles (negative or positive) shall be the victor.

- Resident N decides to backstab resident G and resident G engages in a traditional duel. In this second case, TNZ will always be victorious. For if a Natural backstabs, they sacrifice their particles.

- Resident G decides to backstab resident N and resident N engages in a traditional duel. For the third case, the Naturals will always be victorious. When a Negative backstabs, they willingly gift positive particles to their enemy.

- Both residents decide to backstab the other. The final possibility, when the backstab technique is used by both, results in a victory for the soldier with less particles[†]!

We call for an immediate ban of the backstabbing technique. From here forth, only the traditional dueling shall be permitted. Whether it be by accident or design, we are all Numberfolk! Only when we begin to live together in peace will we flourish as a society.

---

[*] We understand backstabbing today as subtraction. The SAN has banned its use and implores only addition. Should we wish to find 46 − 23, we simply recreate a battle with 46 and 23's evil twin, −23. The result is 46 + (−23), an equivalent duel.

[†] Similar dilemmas still exist today. The well-known *Prisoner's Dilemma* is used by police to attempt to entice suspects to backstab their criminal colleagues.

The Integer Wars, while tragic, allowed Numberfolk a new understanding of themselves and their world. Numberfolk agreed with the Thinkers' declaration and the wars came to end.

Today, the educational system refers to the duels between residents of Natural and TNZ as *addition* and *subtraction*. They force unwitting *children* to recount these horrific battles of the Integer Wars as mere exercises on worksheets filled with happy and pleasing drawings such as snowmen, turkeys, and flowers.

To residents of Numberville, the wars were anything but joyful. They gladly pushed their past away and focused their attention on the next era: The Expansion.

BEHAVIORISM UNIT 1

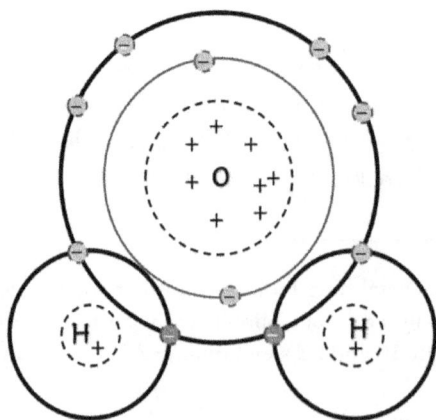

144 − (−24)

Many may wonder about negative numbers. Why do they exist? Now we understand they were the result of the Great Scale ensuring balance. And while negative numbers are used to describe day-to-day concepts like debt, it turns out, they have embedded themselves into *everything*.

The entire Universe is made up of atoms. Everything around you, inside of you, everything you see and touch, are built from atoms. Inside atoms are negative particles – called electrons, and positive particles – called protons.

To appease the Great Scale, the number of protons and electrons are the same, balancing each element. When you look at the Periodic Table, the Atomic Number (the little one on the top left) represents the number of protons.

This *infuriates* TNZ. To this day, they still discuss why the Atomic Number is not defined as the number of electrons. It is, after all, the electrons that do all the hard work.

When substances are created, electrons, the negative particles, are the ones out there negotiating. You see, each element wants to be "full" or "empty" (no in-between). Negatives negotiate to either give away electrons or share electrons so everybody is happy.

In water, H_2O, each H has 1 electron, and the O has 8 electrons. The H's want two each, and the O wants 10. So, they negotiate. They all decide to share. Each H, hydrogen, ends up full with 2 and the O, oxygen, gains the 2 it wants as well. Win-win.

We can see how important the negative and positive particles are, but as a warning, if you ever encounter a resident of TNZ, tell them the Atomic Number is the number of electrons, you'll be instant friends. Whatever you do, don't tell them that their negotiating is exactly why the Atomic Number is actually defined as protons. Since they are out there sharing and giving themselves away is why we need a reliable description like protons, positives, who are all safe and stable inside the nucleus. Whatever you do – don't tell them that.

11

ADDITION

"MATHEMATICS IS THE QUEEN OF SCIENCES AND ARITHMETIC IS HER CROWN."

-GAUSS

It was Sunday afternoon and Marco and Mr. Pikake had already begun in their private study room. The professor spun on one foot to face Marco. Like the ringleader of an invisible circus, he stretched out his arms and announced, "You know the four operations! Addition. Subtraction. Multiplication. Division." Elongating his bony fingers and pointing to each as he recited. "Our first step will be to explore addition and subtraction."

Marco groaned, "I know how to add. I thought we were going to start hunting today?"

"*Ahh*. You have learned to recreate the ancient battles. To become a master, we must study these more deeply."

The boy's rollercoaster was slowly inching upwards, building anticipation for the drop ahead. Marco's blood tingled with excitement. He forecast that today would finally be the day he learned about Numberfolk duels.

Mr. Pikake sat down and leaned over the table. "Marco." He was staring the student directly in his eyes, "I am now going to share with you the key to hunting Numberfolk." He took a deep gulp swallowing his excess saliva. "The key to hunting is to get the enemy *alone*."

A chill ran through Marco. Something about Mr. Pikake's tone made him feel dangerous. Before he could process the feeling, his tutor was back on his feet prancing about the room and the tension faded.

"Now. It may take a significant number of steps to accomplish this task of isolation. Numberfolk work together to hide and distort their actions, their intentions. But!" A pop. "We can use their own history, their own patterns, against them. It is time I explain to you the duels."

In a dramatic reenactment, the tutor described to Marco the history of Numberville from the Fable of Nava through the Integer Wars. He drew intricate pictures on the whiteboard while acting out the duels with his hands, slamming them together to

$$271 - x = 99$$

demonstrate the collision before cupping his palms to indicate something new was created. Marco was a child in a toy store. Suddenly, he forgot he was learning math, this was a complex video game filled with battles and ripe with strategies.

"There are a few key concepts to take from the history of Numberville. The first, is a change in perspective. Your whole life you have learned that addition *and* subtraction are operations. The Numberfolk do not see it this way. For a duel is always what you know as addition. What you have come to know as subtraction is the horrible process of backstabbing. It is unnecessary and clutters the battlefield." He began to scribble on the board.

$$5 - 3 = 5 + (-3).$$

"Thus, we will remove this operation from our repertoire. We never need to subtract while we hunt. Instead, we will force Numberfolk to duel. For subtraction is simply a duel with the evil twin."

Marco had covered integers in the fall before working with his tutor. He felt good about the whole idea because there had been no letters involved, only numbers. The unit in Mrs. Sanders' class had begun with the introduction of negative numbers. Marco liked to think of negative numbers as digging down, while positive numbers were like building up. The positives were easy, they had been working with those since the first grade. Negatives were new and could be tricky. The hours he spent in *Minecraft* helped solidify the concept.

The overall idea of the unit was this: If you add two negatives, you get another negative. If you dig a hole and then dig some more, you just have a deeper hole. The great thing about this was that adding negatives was the *exact* same as adding positives, you just had to slap a negative sign in front.

$$8 + 9 = 17 \ \textit{and} \ -8 + (-9) = -17.$$

$$439 + y = 612$$

The next major topic was what happens when you add a negative and a positive. This was more complicated because sometimes you got a positive and sometimes you got a negative.

$$-8 + 9 = 1 \; but \; 8 + (-9) = -1.$$

How he eventually came to figure things out was this: First, he would ignore the negatives altogether and just subtract the smaller number. If Mrs. Sanders asked him to find $-23 + 15$, he'd just take the two numbers without any negatives, 23 and 15, and subtract the smaller number to find $23 - 15 = 9$. Then, *if* he remembered, he would go back to the problem and add the sign from the first number, the bigger number, to his answer. In this case, he had 23 and 15. Since 23 was bigger, he looked at its sign from the question, which was negative, and then attached that to his answer to get $-23 + 15 = -8$.

This all worked out okay, but as with any math trick, it meant he was following steps without actually understanding what he was doing. This meant if he forgot a step, which at least half the time he did, all was lost. He tried to imagine this same duel in Numberville. He saw each Numberfolk take their ten paces and then run at full speed colliding into each other. The whole thing made him laugh. They broke apart into particles: -23 became twenty-three minuses and 15 became fifteen pluses. Each minus and plus found each other and together formed a star $*$ that exploded, vanquishing each other. Only the eight minuses without a partner remained. He wasn't sure if this line of thought would be easier, but it was certainly more entertaining to see the stars, like fireworks, explode in the air during a test. It also had the added benefit of Marco being less likely to forget going back to check the sign.

$$-43 = c - 217$$

Mr. Pikake was still talking, Marco snapped back to hear, "The second key idea in a duel is what we learned from the Battle of Cancelation. That is, when any Numberfolk duels its evil twin, they vanquish each other, and the result is a zero. But," this pop rang out like a firework, "there is a key part of the story that is not included in the history. That is, the duels with zero."

Marco leaned in, ready for a wonderland of excitement. He already had enough ammunition to keep his imagination busy for days, but now he was offered even more – the Easter eggs. The hidden information only available to those dedicated enough to search.

"Before the Integer Wars, there were duels with Zero for various reasons. The whole thing is a bit of a mystery. Numberfolk would declare some qualm with Zero and announce they were off to duel. The strange thing was, they would always return as themselves – unchanged. At first, many doubted a duel even occurred. They claimed it was just Numberfolk talk 'dueling with Zero'.

As time went on, arguments erupted over the matter. Some thought Zero refused to duel, they branded him a coward, a deserter, saying he always ran from a fight. Others believed it was magic. They thought Zero, being born from Zil, possessed a reflective coat. The important thing for hunters is that any Numberfolk, be it a Natural or otherwise, who duels with Zero always leaves the battle just as they entered – completely unharmed."

"Did the Thinkers determine what it was?" Marco asked eagerly.

The professor chuckled, "Excellent question. The SAN doesn't have any documentation of such a query. You can decide for yourself what to believe."

The idea of a magical cloak of reflection was too much for Marco to pass up. Wanting to document these ideas, he snatched

$$-78 + x = 97$$

his journal from his backpack. Across the top of the next blank page, he wrote:

Hunting 101

before wrenching his neck backwards to look directly at Mr. Pikake indicating he was ready and listening.

"Wonderful! Today, we are concerned only with Numberfolk who camouflage themselves with duels. Our goal will be to vanquish any of their supporters, those helping the numbermask to hide. To successfully stalk these creatures, we have three techniques. The first is to force a duel. While we can enact a duel with any Numberfolk, we shall be most successful using technique #2, sending in the evil twin. When we force a duel between a Numberfolk and its evil twin, they will always vanquish each other. We may use techniques 1 and 2 as many times as needed until the numbermask is isolated, alone. Our final and third technique is that once we have stripped the sympathizers away, the numbermask has only a duel with Zero remaining. Thus, the mask is alone, and we are able to finish it off – complete the hunt."

Marco diligently jotted down his professor's words.

1. Force a duel.

2. A duel with an evil twin results in zero.

3. Require the mask to duel with zero.

After dotting his last 'i', Marco enthusiastically called out, "I got it!"

"Well, then." A dark grin spread across Mr. Pikake's face. "Let's hunt."

Sitting on opposite sides of the table, the two leaned in – only a piece of paper between them. "A hunt is only possible with an equation. That is, we know the pattern and we know the result. Only then can we identify who is hiding."

$$413 = 589 - x$$

Marco thought about this for a moment until he felt convinced it made sense. He examined his factory drawing on the adjacent page. *If there was only an expression, a pattern, then things are always changing. With* $20 - x$ *if* 9 *goes in, then we see* 11, *if* 7 *goes in, we see* 13. *There is not a lone Numberfolk hiding. But if* $20 - x = 13$, *then it must be* 7 *hiding.*

"We will manipulate, maim, and eliminate numbers in the hunt. These are all distractions to help the concealed. Don't become overzealous! You may be tempted to try this or try that which can lead you down the wrong path. This has happened many times, often with devastating consequences. You must maintain the basic truths."

"But how do I do that?" Marco interjected.

"Proficient probe, lad." Mr. Pikake continued. "Truth is only achieved through the perfect balance of a scale. To maintain said balance, if you take from one plate, you must also take from the other. If you give to one plate, you must equally give to the other. Only then can stability settle."

Marco was bewildered. Balance, scales, duels, reflective coats, and hunts. It was a lot for him to keep track of. But to Marco, it was wonderful. He imagined himself as his avatar preparing for a quest. He looked up at his tutor, "Let's go."

"First, a simple hunt. Then, we may tackle more advanced forms of obstruction. If I asked you what duels with 7 to result in 12, could you tell me the partner in this challenge?"

Decoding the words, Marco understood. He was being asked to determine what number when added to 7 is equal to 12. "Five." Marco spoke quickly and surely, just as when Liam or Oliver questioned him during one of their campaigns. Speed and accuracy were always important on any quest.

"Very good." The quirky playfulness his tutor normally displayed had vanished. The two spoke as if military men discussing their offensive. Mr. Pikake scribbled on the paper between them. "We may represent this as…"

$$a + 7 = 12.$$

"Now, *we* know 5 is hiding behind a, which makes this an exemplary example to begin with. There is a dance we can do with any equation in this form that *always* results in our victory."

The familiar 5 stuck its head out from behind a. Giggling, it took 7's hand and they began to dance. Marco pushed away the daydream, he was determined to focus.

"What did I tell you our goal was, Marco?"

"To get the numbermask alone. To vanquish its supporters." Marco caught himself wanting to firmly end with 'sir' but stopped before the word slipped out.

"Perfection. And who is the only Numberfolk who may duel with whomever is obscured behind a and result in a?"

This one was a bit harder. Marco reminded himself that dueling was addition. So, the question was really, what plus a is equal to a. "But we don't know who is hiding. How can we know who it should duel with if we don't know who *it* is?"

"*Ah*. Remember there is but one Numberfolk who no matter whom they duel with, the result is always their partner. This is true no matter who may be trying to trick us."

"Oh! Zero. His reflective coat means that in a duel with a, the result will be a. Or $0 + a = a$ no matter who a is."

"I see you have chosen your school of thought." Mr. Pikake chuckled noticing Marco had decided to see Zero as magical rather than the opposing option of a coward. "In order to isolate a, to get it alone, we want it to ultimately duel with Zero. Once that duel is completed, all that will be left standing is a."

"Okay, okay," Marco nodded. He examined the $a + 7 = 12$ on the paper. "But a is *not* dueling zero, it's dueling 7."

"Keen observation, lad. That brings us to the first step in our dance. We will force 7 into a duel. If the result of this duel is 0, then we are done."

"We can do that? That seems like... like... *cheating*," Marco responded wearily.

"High morals. I admire that. As long as we maintain the balance, it will remain fair. That is, if we take from one plate, we must take equally from the other. If we force 7 into a duel, we must also force the same duel onto the result of 12. But who should we send in?"

Clearly Marco was meant to answer this question, but he didn't have a clue. He was still trying to regain his own balance to fully understand the hunt.

"Stay focused on the end game, son. Remember, we ultimately want a to battle 0. Whomever we send in to battle 7 will work in our favor if the result of that duel is 0."

"Its evil twin!" Marco shouted throwing his hand up in victory.

"Exactly! We send in the evil twin to duel. When the twins battle, the result is zero. This vanquishes the supporters and leaves the mask to duel with zero ultimately isolating it. Game. Set. Match." Mr. Pikake leaned back in his seat and shot Marco a look of contentment.

"Okay. I think I am getting this. But what happened to 5? Where does that come in?"

"What you have just developed is an *algorithm*. While Numberfolk have the magic of the Spogs, humans have algorithms. They are our own magical spells that we can cast to reveal any Numberfolk evading us. In cases like these, we complete three steps." He raised his three long fingers and said, "One. Send in the evil twin to duel leaving a zero in their wake. Two. The numbermask must then duel with zero. Three. The numbermask is alone, we complete the hunt." He motioned to the paper between them. He repeated himself, "One. Send in the evil twin to duel. Who do we send in?"

$$a + 7 = 12.$$

"We send in −7, that is the evil twin of 7," Marco pointed to the sheet.

"Exactly. But we must maintain the scales, we send −7 to battle both sides." Mr. Pikake added that to the paper.

$$a + 7 + (−7) = 12 + (−7).$$

"Now what?"

Marco studied the equation. "The duel on the left results in zero. $7 + (−7) = 0$. The duel on the right results in 5." He paused. "Oh! I see it now. The duel on the right results in 5 because $12 + (−7) = 5$." Using his new concoction, Marco imagined twelve pluses and seven minuses in the sky. Seven pairs created exploding fireworks above the 5 pluses that remained.

Mr. Pikake transcribed what Marco had described.

$$a + 0 = 5.$$

"Two. The numbermask duels with zero. What is the result?"

"a! The result is a! Zero's reflective coat leaves a alone, exposed!" Marco was thrilled. The professor continued as he recounted the final step, "Three. The numbermask is alone, we complete the hunt."

$$a = 5.$$

As complex and perverted as it sounded, the whole thing was starting to make perfect sense to Marco. He imagined the battles that had taken place. A warrior approached the base of a volcano where 7 lay slain on the ground. As he scanned the scene, he saw the goddess of the volcano holding a scale that read 12. The warrior asked the goddess, "Who has battled 7 to cause this destruction?"

$$a − 90 = 90$$

The goddess responded, "For only one exists who could harm 7 in this way, t'was his twin." The warrior, knowing that −7 had ravaged his brother looked to the scales. He slid the measurement and understood who was there. Knowing 5 hid and allowed this atrocity to occur, he set forth to find and punish it.

There was one thing Marco wanted to know. "Does the al-gor-thm always work?"

"It does! The algorithm will be successful anytime we have a mask using duels to conceal itself. In both simple and more complex concealments." Mr. Pikake looked directly at Marco and in a soft voice asked, "Do you think you are ready to hunt solo?"

Marco nodded. Not only was he excited to try this out, he also desperately wanted to prove to his tutor that he could be successful. That it was not a mistake to put his faith in Marco, to gift him the journal.

"Wonderful." The professor presented Marco with his hunt before sitting back and motioning for the student to take over.

$$a + 12 = 20.$$

Marco rubbed the palms of his hands together. "Okay." He started, "First, I send in 12's evil twin to duel. When 12 duels −12, the result is 0 and when 20 duels −12 the result is 8." He wrote his steps.

$$a + 12 + (-12) = 20 + (-12)$$
$$a + 0 = 8.$$

"Now I've forced a to duel with 0, the result is a," he looked at his tutor and added, "because of Zero's reflective coat." Mr. Pikake gave Marco a wink and grin. "That means it is 8 hiding behind a!" Marco jumped up and thrust his arm forward stabbing an invisible force.

"Remarkable, son! Simply remarkable." Mr. Pikake clapped his hands and stared adoringly at the boy. With a mischievous grin he asked, "Care for another?"

$$x - 9(8 + 1) = 100$$

"Yes, please!" Marco responded enthusiastically.

$$c - 5 = 15.$$

This is a deceiving situation. Mr. Pikake explained. "But," he popped, "I believe you are prepared for the challenge.

"Will the algorithm still work?" Marco asked.

"Yes! Complete the same spell work, the same dance. You are capable."

"Alright. I send in the evil twin first. I force -5 to duel."

$$c - 5 + (-5) = 15 + (-5).$$

As Marco continued, he realized he had a problem. "Wait. $-5 + (-5) = -10$. Which is *not* zero. Which means c is *not* alone. Which ruins everything."

"Excellent observation. What was your mistake?"

Marco wasn't sure. Then he remembered the very first thing Mr. Pikake had told him. "You said we don't need to subtract anymore. If I change everything to a duel... then..." He erased his previous markings and replaced the problem.

$$c + (-5) = 15.$$

"Now I see it. It is -5 that is dueling with c, not 5. I should send in the evil twin of -5, which is 5 instead."

$$c + (-5) + 5 = 15 + 5.$$

"Then, when -5 and 5 duel, the result is zero. And, when 15 and 5 duel, the result is 20."

$$c + 0 = 20.$$

"Now, c will duel zero which leaves c alone and I know it's 20 that is hiding." He looked to his tutor for approval. Mr. Pikake nodded and smiled. Marco clinched his fist and threw his elbow down in a moment of victory.

$$321 = x + 139$$

"You have done extraordinarily well today, son. You should be proud. You have mastered what some call the one-step hunts with duels. They call it this since we need only to send in a single Numberfolk, the evil twin. I must warn you…hunts are rarely this clear. From my observations, I believe you are ready for the next challenge."

"Thank you, thank you, professor," Marco spurted. "I think I am ready, too."

"Attempt this advance."

$$3x + 6 - 2x - 9 = 24.$$

Just looking at the equation made Marco feel dizzy. Out of the corner of his eye, peered his werewolf sketch in his journal. He had an idea. He wanted to try to group the werewolves together, to get a better idea of what he was up against. Not wanting to get confused, he first changed everything to a duel.

$$3x + 6 + (-2x) + (-9) = 24.$$

Next, he grouped the werewolves and the not-werewolves together.

$$3x + (-2x) + 6 + (-9) = 24.$$

I have only one werewolf here, but who are they dueling? He imagined 9 minuses and 6 pluses. After the 6 fireworks exploded in the sky, he saw there were 3 minuses remaining.

$$x + (-3) = 24.$$

Phew. Now this looks like a normal hunt, he thought. Speaking aloud he said, "I send in 3 to duel its twin."

$$x + (-3) + 3 = 24 + 3.$$

"The twins cancel, leaving zero and the duel between 24 and 3 results in 27."

$$x + 0 = 27.$$

$$277 - x = 94$$

He saw the finish line. In his excitement, words rapidly flew from his mouth, "When x duels with zero, the reflective coat shows just x, so..."

$$x = 27.$$

Marco slashed the marks across the page like an artist with a paintbrush.

"And you doubt your greatness?!" Mr. Pikake finally spoke. "Magic, pure magic, son. Well done!"

The boy filled with pride like an overstuffed bear, nearly bursting at the seams.

"One final hunt for today, I find myself exhausted!" The professor wrote the last equation on the paper.

$$24 - 2y = 9 + (-3y).$$

He has to be kidding! They're everywhere! Marco thought as he noticed the y-mask on both sides of the equation. Hesitantly, he decided to vanquish the 24 to see where that got him.

"*Ahem. Um.* I'll send in 24's evil twin to force the duel." He slowly wrote the step hoping his procrastination would give his brain time to come up with a better idea.

$$(-24) + 24 + (-2y) = (-24) + 9 + (-3y).$$

"So, *ah*. The duel between -24 and 24 gives us zero. And the duel between -24 and 9 gives us..." His brain was overwhelmed. There was too much going on. He had the -24 spit out four negatives and the 9 spit out four positives to result in $-20 + 5$ watching the fireworks explode. This was enough to get him back on track. "It gives us -15 which means..."

$$0 + (-2y) = -15 + (-3y).$$

Still trying to buy time he continued with, "Zero's reflective coat means $0 + (-2y)$ is just $-2y$."

$$-2y = -15 + (-3y).$$

$$367 + y = 551$$

And that was it. He was as far as he could get. Marco imagined he was being chased by numbermasks. He turned down a dark alley, bad choice. He was stuck with no way out. They were surrounding him, closing in. If he could just herd them together, he could take them out in one move.

"What is our goal, son?" Mr. Pikake spoke softly.

"To get the numbermask alone," Marco sighed.

"You still have masks everywhere, don't allow them to win!"

"I know, but I don't get how to get all the masks together."

"Well, we need to vanquish one of the masks...to get rid of it."

Marco stared at what remained of his hunt. He wanted nothing more than to get rid of the masks. They sneered at him, taunted him, they were winning. "But how can I do that?"

"How would you normally vanquish a term?" Mr. Pikake was clearly leading him down a road, but Marco was struggling to see the path in front of him.

"I force a duel with the evil twin."

"Perfect! Do that!" Mr. Pikake excitedly pointed to the paper.

"I don't know who $-3y$'s evil twin is though. Because I don't know who y is."

"*Ah*! That is the exact question we must ask ourselves. Who is the twin of a mask?" He jumped up and went to the board. He slowly drew out the letter b. "Who is b's twin?"

"I'm not sure, it depends on who b is!" Marco groaned.

"You must fight a mask *with* a mask, son. You know that when this b duels with its twin, the result will be..."

"Zero. Anyone dueling with their twin cancels out."

"Good! So, who shall duel with b?"

$$22 = b - 163$$

Like a puzzle, Marco started trying to put the professor's words together in a way that made sense. *I need to fight a mask with a mask and find who will duel with b to get zero.* After a long pause Marco mustered out what was at best a guess, a hail Mary, "With $-b$?"

The tutor jumped to life, "Superb son, superb!" He scribbled:

$$b + (-b) = 0.$$

"Now, back to your hunt. You wish to vanquish $-3y$. Who is their twin?"

Unsure if luck was on his side or if he was onto something Marco blurted, "$3y$!"

"Exactly!"

"Okay. I send in $3y$ to duel on both sides."

$$-2y + 3y = -15 + (-3y) + (3y).$$

"Three y's minus two y's leaves me with just a y!" He saw it, he knew who he was hunting. He knew what to do. He saw a side door in the alley just in time and burst through. As the masks pursued, they were pushed together into a single area allowing Marco to vanquish left and right. He would arise victorious!

$$y = -15 + 0$$
$$y = -15.$$

Mr. Pikake bowed, "Amazing work, young master." When he rose, a smile began to grow across his face.

Marco felt strong. He felt powerful. Not only did he understand the duels, he loved them. He adored Zero's reflective cloak, he was obsessed with vanquishing. His brain felt like it grew two sizes and was nearly exploding out of his head.

Allowing the student time to let his supremacy soak in before continuing, Mr. Pikake gently added, "Since I will not see you until next weekend, I shall leave you with a small homework assignment to keep your skills sharp until we see each other again." He gathered his things, slipped on a thick overcoat and a

$$-84 + m = 102$$

bowler hat before handing Marco a slip of paper. Tipping his hat to say goodbye, Mr. Pikake slipped out of the room.

Marco was on a high. He pretended to hunt and fight imaginary enemies his entire way home. It had been a long time since he felt this good. He was becoming a warrior. To the outside world, some might say he was simply beginning to look like 'a math kid'.

As he turned the corner, he caught a quick glance of the speed limit sign. The 35 seemed inconspicuous enough. It stood tall like a crossing guard stoically performing its duty. He tipped his head as a show of respect to the number. As he rose, directly behind the sign, he saw another…an out of control 47 speeding towards him. A cowboy on a steed, it rode upon a beat-up truck laughing hysterically as it grew closer. Marco recognized the truck, it was Peter's. He took two long strides backwards, too scared to tear his eyes away from his stepfather or the number. The heal of his shoe caught a crack in the sidewalk which pulled him down to the hard ground. He felt the sidewalk's icy breath beneath him.

The brakes squealed loudly as Peter punched to a stop. Before the vehicle had the chance to jerk back from the sudden shift in momentum, Peter had already jumped out and was racing towards Marco. With a quick jerk, he grabbed his stepson by the collar of his jacket, lifting him from the earth. "You aren't so tough all alone, are you?" He spit in the boy's face as the words hurled violently from his mouth. As quickly as Peter appeared, he was gone. Marco dropped back to the sidewalk in a daze of confusion.

He blinked. Was he imagining this entire event? Marco looked to see Peter's truck, still standing at an angle, half in the street, half on the sidewalk. No. It had really happened. Where did Peter go? Marco twisted to his knees to find Peter laying on the ground next to him. Scuffed black dress shoes and brown creased slacks came into his field of vision. He tilted his head slowly upward to see who they belonged to. The friendly familiar face of his tutor towered over the two.

$$312 = 499 - x$$

"What the hell?!" Peter sprung up, ready to fight. Without a word Mr. Pikake scooped Peter by his shirt and pushed him against the speed limit sign. If Peter had the thought of trying to fight back, it wouldn't have gotten him very far. Mr. Pikake's long arms created an invisible barrier between the two. Peter could swing and swing and never once make contact with the professor.

"Don't you *ever* lay a hand on that boy. I'll have you in jail faster than you can blink." Marco's tutor said in a bellowing voice, his perfect punctuation as clear as ever, before releasing his grip causing Peter to crumble to the ground.

Helping Marco up, Mr. Pikake ushered his student towards home. Marco struggled to keep up with the professor's long, blithe paces though Mr. Pikake's hand firmly pushed the boy forward. As he tried to gain his bearings and come to grips with the event that had just taken place, he looked back. He was surprised Mr. Pikake was brave enough to turn his back on Peter. He had upset a rabid animal who didn't have the pride to resist an attack from behind, an opponent who would readily backstab his prey. Surprisingly, Marco saw his stepfather still on his knees beneath the sign.

Mr. Pikake dropped Marco at home. He didn't say a word, he simply pushed the boy inside before turning and walking back down the driveway. Marco ran to his room and slammed the door. He was in a full panic. The professor had stopped the attack, but Peter lived here. He would be coming home any minute and who would protect Marco now? He ran to his window monitoring for the truck. As it slowly pulled in front of their house, he noticed Mr. Pikake hadn't left at all. He stood under the oak tree across the street, his lean body rested on the trunk like a ladder. Peter must have seen him, too. He kept looking over his shoulder as he slammed the truck door and made his way to the porch.

"MARYANNE!" Peter's shouting rattled the entire house. The next hour was filled with yelling. His stepfather screamed about how the tutoring was over. How Mr. Pikake was a crazy man who attacked him for no reason. Eventually, Marco was called down. He explained that the professor was only trying to protect him. Incredibly torn, Marco didn't want his mother to be upset but he didn't want to lie either. He was ultimately sent to his room while the two adults battled on. That night, his mother brought Marco his dinner upstairs.

"I figured everyone needed some time to cool off." She said quietly. Placing the tray on his desk, she sat next to Marco on the bed and began to cry. His heart tore apart. He gripped his mother tightly hoping his sheer force would stop the sobs. "Why can't you get along with him?" she wailed. "I know he has his temper and hates to not get his way…but, he takes care of us. He tries."

Marco was dumbfounded. How was this his fault? "I-I" he had no words.

She sniffled and wiped her face with her sleeve before standing back up. "I don't know what really happened. Peter is infuriated that I took your side. I tried to fix it, I said we'd fire Mr. Pikake and find you another tutor."

Marco felt like the air was sucked from the room, he couldn't breathe. He didn't want another tutor. He started to speak up but his mother kept going, "That didn't make him happy either. He doesn't want any tutor, which seems ridiculous. Just let him cool off, avoid him for a while. He will have to get over it." She turned and left the room, softly closing the door behind her.

His rollercoaster was twisted in knots like his stomach. He pushed the food away and lay in his bed with the blankets tightly wrapped around him. Trying to think, trying to plan, his ideas were like chickens running around. Every time he caught one, it slipped away again. Finally, he gave up and allowed his emotions to flood over him. Tears rushed down his face until sleep won and took him under.

HUNTING UNIT 1

$$x - 126 = 63$$

$3 + 4 + 5$

$3\sqrt{16}$

$\dfrac{6}{3}$
$\dfrac{}{6}$

$\dfrac{1}{2}(4!)$

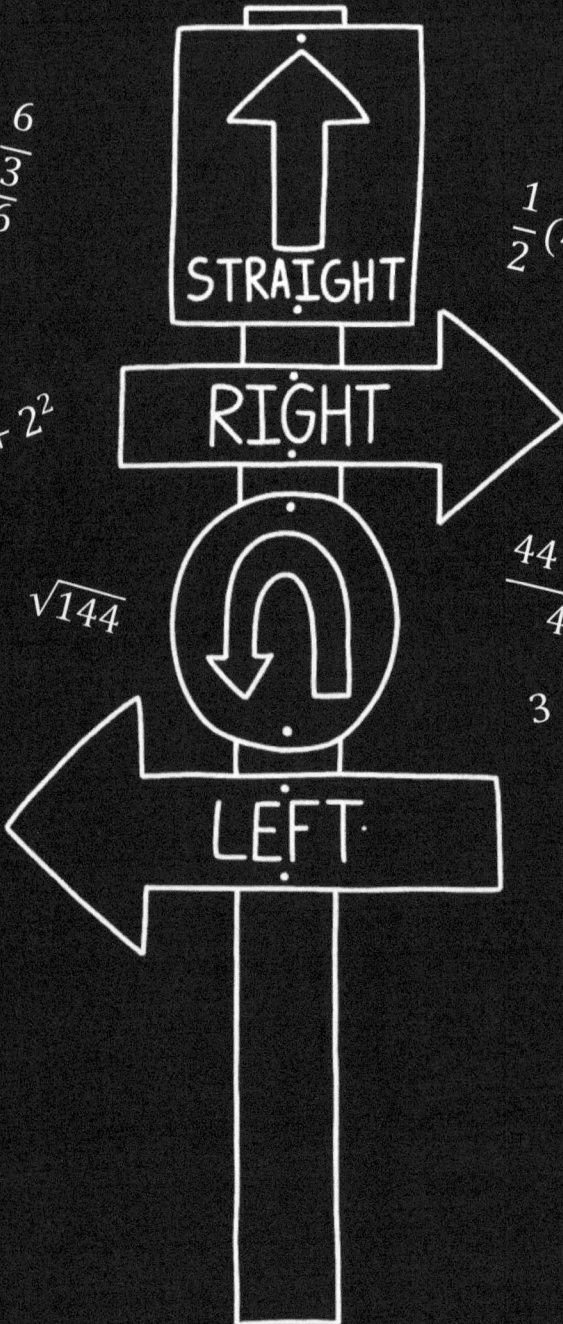

STRAIGHT

RIGHT

$2^3 + 2^2$

$\sqrt{144}$

$\dfrac{44 + 4}{4}$

$3 \cdot 2^2$

LEFT

12

HOMEWORK

*"THE ONLY WAY TO LEARN
MATHEMATICS IS TO DO
MATHEMATICS."*

-PAUL R. HALMOS

W hatcha doing?" Maggie was bopping up and down and rolling her neck back and forth to silent music.

"Homework," Marco said dryly, not looking up.

It was Monday night and Marco was ready to begin his hunting practice. Peter had been avoiding him like Covid. They hadn't been in the same room since Sunday night's explosion. Marco had felt weak and helpless for a full eighteen hours before he decided to stop moping and do something about it. This had to be worth it. He was going to become a master, he was going to make his mother happy, and he was certainly going to become strong enough to protect her and Maggie.

He opened the slip of paper and unfolded it on his desk. Disappointed, the sheet looked like any math worksheet. Well, it *almost* looked like any other worksheet. It was significantly more boring. He wasn't sure what he expected – dueling Numberfolk lining the margins? All that was on the piece of paper were the weekdays followed by three equations for each day. Marco thought he understood: they couldn't have papers laying around talking about hunting and Numberfolk, so Mr. Pikake voted for simplicity.

He looked at the first problem.

$$x - 5 = 20.$$

"That's easy. It's 25." Maggie interjected cheerily.

"How'd you know that?" Marco was both impressed and annoyed. This was Marco's special world. Leave it to his sister to beat him at his own game.

"Because the only thing you can subtract 5 from to get 20 is 25. So x has to be 25."

Marco remembered his first hunt with Mr. Pikake. They used the same reasoning. This was another good opportunity to try out his algorithm spell. He first got rid of any subtraction so that everything was written as a duel.

$$3y + 60 = 2y + 252$$

$$x + (-5) = 20.$$

He then sent in the evil twin of -5, five, to duel both sides.

$$x + (-5) + 5 = 20 + 5.$$

When -5 dueled 5 the result was 0, and when 20 dueled 5 the result was 25.

$$x + 0 = 25.$$

Lastly, x dueled 0. Marco imagined x approaching Zero. Suddenly, Zero whipped his arms around and vanished. As x approached the spot where Zero once stood, there was only a mirror and x saw his own reflection. He snickered.

$$x = 25.$$

Maggie looked down at the paper. Her eyes doubled in size and she shrieked, "Teach me!" His sister was smart, but her class wasn't working on number-letter math yet. Marco was sure she just wanted the ability to brag to everyone about how much smarter she was. He also knew he was not going to get out of this easily. His sister wouldn't give up until Marco did as she wished and worse, Marco didn't think Mr. Pikake would want him to reveal the Numberfolk and the duels, so he didn't even know *how* he could teach her without teaching her.

He rubbed his eyes and thought hard. He decided to teach her the algorithm in boring mathy terms – she'd love that anyway. "Okay. The first thing you need to know is that in harder math they don't use the minus, times, and divide anymore really. Instead of subtraction we just add a negative, and instead of multiplication they just use a dot or parentheses. *Unless* there are letters, then they don't use anything at all, and..." Marco realized he hadn't really done much with fractions and wasn't exactly sure how it worked. "*Um.* Well, division, they just write as fractions instead."

"WHAT?!" Maggie burst out. "Are you kidding me? *Ugh.* Leave it to the teachers to do something crazy like this. This is just like the Cairo thing. Do you remember that? Why are they

$$x + 64 - 2x = -129$$

teaching us all this wasted knowledge to just change it later? I am actually appalled." Marco couldn't help but laugh. He felt the same way when he first heard about the notation changes but admittedly for different reasons.

"I thought that, too." He shot his sister a kind smile before continuing. "Anyway, on problems like this, all you have to do is get the letter by itself. But anything you do to one side you have to do to the other." He held his arms out, each palm facing up before tilting to indicate an unbalanced scale. Grabbing a fresh sheet of paper, he wrote a new question down for Maggie.

$$9 + y = 14.$$

"Since you have 9 plus y, you need to get rid of the 9, which you can do by adding -9."

$$-9 + 9 + y = -9 + 14.$$

"Now -9 plus $9 = 0$ and -9 plus $14 = 5$."

$$0 + y = 5.$$

"I can get it from here," Maggie began. "Anything plus zero is just what you started with, so $y = 5$. *And* that makes sense because the number you add to 9 to get 14 is 5. Alright. I'm ready for a harder one."

Of course, Maggie would pick up on everything with ridiculous speed, Marco thought. He didn't have anything harder to give her. He wouldn't even know where to begin. He instead gave her a playful shrug and told her to look it up online and leave him alone.

By himself, Marco focused on the next Monday problem.

$$4y - 7 - 3y + 9 = 42.$$

He gulped, regretting having sent his sister away. *Trust the algorithm. This is a messy situation, made to confuse me, but it is just a Numberfolk hiding. I can do this.* Leaning on the motivation from his personal pep talk, he began. First, he didn't need the subtraction to confuse him, so he turned everything into a duel.

$$9c + 15 = -c + (9c + 209)$$

$$4y + (-7) + (-3y) + 9 = 42.$$

Marco knew he could rearrange terms[*]. The reason was because $3 + 4$ is the same as $4 + 3$. However, Marco also knew that the Order of Operations he struggled with at the start of the year was important. He stared at the paper and convinced himself that since they were all duels, all addition, it was okay to rearrange. He wanted to make the problem look more like the ones he had solved before. So, he put all the masks together and all the numbers together.

$$4y + (-3y) + (-7) + 9 = 42.$$

He knew when -7 dueled 9, the result was 2. But, what would happen when $4y$ dueled $-3y$? A werewolf roared ferociously, popping out of the desk, trying to take a bite out of Marco. It was his mind's way of connecting the dots. "Like terms!" he said aloud. He imagined his werewolf/vampire soiree. Four vampires stood in the corner, one of them said something offensive and the other three walked away leaving the lone vamp.

"If I have four y's and I take away three y's, I would be left with one vampire, er, one y." He chuckled to himself.

$$y + 2 = 42.$$

Marco was amazed. He knew who was hiding, it was 40. Not only that, he had managed to navigate this new unchartered territory to get to something that he knew just what to do with. AND, the whole algorithm was easy for him now. He wrote the last steps in a few quick strokes of his pencil. (He always did math in pencil, easier to handle when he made a mistake.)

$$y + 2 + (-2) = 42 + (-2).$$

---

[*] This is called the commutative property of addition. Multiplication is also commutative. Terms being summed may be rearranged. $1 + 2 + 3 + 4 + 5 + 6 + 7 + 8 + 9 + 10 = (1 + 9) + (2 + 8) + (3 + 7) + (4 + 6) + 10 + 5 = 10 + 10 + 10 + 10 + 10 + 5 = 55$.

$$4k - 9 - 3k - 16 = 170$$

$$y + 0 = 40$$
$$y = 40.$$

I literally can't even believe I just did that. He looked over his hunt with pride. He was even able to hold onto the good feeling for a full forty-five seconds before the ugly self-doubt monster crept in. *What if this isn't right? Maybe I can't rearrange that way?* The last thing in the world Marco wanted was to present Mr. Pikake with his homework to find out, like his tests, he was a failure. It was the same feeling he had back in the fall when he was so sure the answer was 30, just to find out he had mis-used the rules yet again.

Wanting to be sure his answer was right, his eyes burned into the paper, looking at the problem until everything became blurry. Finally, the numbers and letters shook on the page, dancing around. The $y = 40$ became a number-jail. The 40 was caught – trapped. Then, Marco saw it. He had documented his entire hunt! As he looked up the page, 40 was still free in the previous steps. He saw it dash behind the y out of sight. *Letters are just numbermasks, so whoever is hiding was always hiding!*

Marco remembered Mr. Pikake had told him that substitution was one of the hunter's greatest powers. *I can see if 40 works at the beginning! Then I will know that I trapped the right Numberfolk!*

$$4y + (-7) + (-3y) + 9 = 42.$$

If 40 is hiding behind y then...

$$4(40) + (-7) + (-3)(40) + 9 = 42.$$

He began to compute. *Four times forty is one hundred and sixty. Subtract seven from that to get one hundred and fifty-three.*

$$160 + (-7) + (-3)(40) + 9 = 42$$
$$153 + (-3)(40) + 9 = 42.$$

Now negative three times forty is negative one hundred and twenty and one hundred and fifty-three minus one hundred twenty is... is 33!

$$153 + (-120) + 9 = 42$$

$$-8m + 9m - 65 = 131$$

$$33 + 9 = 42.$$

And thirty-three plus nine is…it's forty-two! Yeeeeessss!

Marco leapt up and began a celebratory dance around his room.

"STOP IT!!!!" he heard the muffled screams of his sister from below. Marco's room was directly above the office.

When they were little, Maggie and Marco thought their mother had a sixth sense as she always knew when they were not laying down past bedtime. They eventually figured out their loud thumping could be heard from below and were more than a little disappointed their mother didn't have supernatural powers.

After dramatically stomping a few more times just to annoy her, Marco returned to his desk. *One more to go.*

$$21 = 15 - x.$$

This was new. Marco didn't remember ever seeing the numbermask on the right side. Trying to reason with the problem first, a horrible feeling crept into his stomach. *This doesn't make any sense! How can I take away something from 15 and get the bigger 21?* He searched his brain for a solution. It was getting late, and every ounce of Marco's energy had abandoned him. His best bet was to try to make this hunt look like the others, then he'd know what to do.

He took a leap. Deciding to try to vanquish the x, he sent in its evil twin to duel. He had no idea *who* this twin was but convinced himself that it didn't matter. He needed to fight a mask with a mask.

$$21 + x = 15 + (-x) + x$$
$$21 + x = 15 + 0.$$

The next part was easy, a duel between 15 and 0 would result in 15. Plus, Marco now recognized things. He sent in 21's evil twin next.

$$21 + x = 15$$

$$12c - 99 + 42 - 11c = 140$$

$$-21 + 21 + x = -21 + 15.$$

Marco saw a path to the finish line, the duel between -21 and 21 was clearly zero and the duel between -21 and 15 would be -6.

$$0 + x = -6$$
$$x = -6.$$

Attempting to check his work, Marco came up with something that scrambled his brain.

$$21 = 15 - (-6)?$$

If $--6 = +6$, all would make sense, 15 plus 6 *was* 21. His paper didn't have 6 though, and he was beginning to understand why Numberfolk didn't like subtraction. With no other option, he opened his mouth and let out a loud cry, "MAGGGGIIIIE!"

He heard her thumping up the stairs before she burst into his room. "Yes?" she said calmly. Putting on her best flight attendant voice she followed up with, "How may I assist you?" before breaking out into senseless laughter. "What's up?"

"What is negative negative six?"

Not sure what her brother meant, Maggie bent over his desk to look at the paper. "Oh, you mean what happens when you minus a negative?" Maggie had participated in some sort of nerd-camp last summer. The focus of the camp was Number Theory. Marco not only had no idea what Number Theory was, he also struggled to understand why anyone would want to learn about it, especially over summer break. Now, looking back, he could see how all of Mr. Pikake's talk about understanding the enemy would make a course like Number Theory have some purpose.

She drew out a long line on the scratch paper. "Here is how they explained it to us," she began. Making vertical marks across the line, she added the number 15 on the far right. "This is the number line. When we add, we move to the right and when we subtract, we move to the left. So, like $15 - 6$ means you start at 15 and go

$$13 + x + 42 = 2x - 143$$

six marks to the left to land on 9, which is why $15 - 6 = 9$." She counted six lines to the left and wrote the number 9 beneath it.

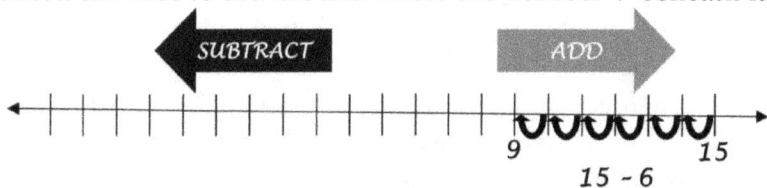

"Okay, but what does that have to do with fifteen minus negative six?" Marco jumped in.

"Let me get there!" his sister responded in an impatient growl.

"A negative or positive tells you the direction. A positive means keep going and a negative means switch directions." Her next words came with an over-the-top whine as if to say 'shut up and just listen.' "For example, $15 + 5$ is plus so you go to the right. And since 5 is positive, you keep going right 5 to get to 20. And $15 - 5$ is a minus, so you go to the left and since the 5 is positive, you keep going left. But, $15 + (-5)$ says to go to the right, but since 5 is negative you actually switch directions and go to the left to get to 10."

Marco remembered why he didn't much care for math. This was unnecessarily complicated and to him, only sounded like a bunch of rules to memorize: go left, go right, no go left, without any reasoning behind them at all.

"So, that means $15 - -6$ is a minus, so it says go left, but since you are minusing a negative number, the negative

$15 - -6 = 6$ *to the right of* $15 = 21$

Go left   Turn around

tells you to switch directions, so you actually go right which means $15 - -6$ is really $15 + 6$. Make sense?" She doodled a little figure to demonstrate her left-right reasoning.

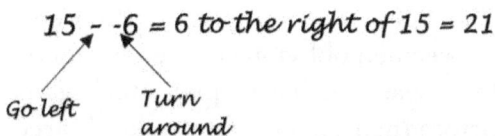

No. It does not make any sense at all. Marco imagined 15 waiting at a stoplight. Instead of the normal red, yellow, and green, the light

$$(3k - 15) - (2k + 18) = 166$$

had two parts. The first was a plus and a minus. The minus shone bright telling the 15 to turn to the left, so she did. Then the second light blinked to life. It was a −6. A big red U-turn sign also appeared indicating 15 needed to turn around. She hopped to face the right, then continued six paces ending up at 21. Although he was still not sure of these rules, he quite liked the game. He played a few more times with new values. 12 − 6 turned left, then kept going forward to 6 while −10 + (−3) started facing the right. When the U-turn sign popped up, they made a dramatic roll, upset from the inconvenience of it all to ultimately take three paces to the left and secure their spot at −13.

"Earth to Marco!" Maggie moaned directly into her brother's ear.

Satisfied with her help, he smiled and nodded, "Thanks Bug," before announcing he was beat and going to bed.

Whatever Maggie had rambled on about was enough to convince Marco his answer was right. If 15 − −6 equaled 21, then −6 must be the one hiding. While the stop sign game was fun, it wasn't much help to understand why 15 − −6 was 21. Making a mental note to ask Mr. Pikake about this next weekend, he headed to bed. So exhausted, before his head had even hit the pillow, Marco had already fallen asleep.

❊ ❊ ❊

The week went by quickly. No one but Maggie spoke at meals, she seemed oblivious to the tension that filled the air. His baseball team was starting to play mock games – it was only two weeks before their opener. Every day Marco came home, ate dinner, did his normal homework, and then, completely drained and dying to sleep, he pushed himself to complete his three hunts before dozing off and doing it all over again. On Friday night, before diving in to finish his assignment, he looked over all he had done.

$$42 + x + x + x = 2x - 80 + 2x - 78$$

Monday

1. $x - 5 = 20$
2. $4y - 7 - 3y + 9 = 42$
3. $21 = 15 - x$

$$x - 5 = 20$$
$$x + (-5) = 20$$
$$x + (-5) + 5 = 20 + 5$$
$$x + 0 = 25$$
$$x = 25$$

$$4y - 7 - 3y + 9 = 42$$
$$4y + (-7) + (-3y) \cdot 1 = 42$$
$$4y + (-3y) + (-7) + 9 = 42$$
$$y + 2 = 42$$
$$y + 2 + (-2) = 42 + (-2)$$
$$y + 0 = 40$$
$$y = 40$$

$$21 = 15 - x$$
$$21 = 15 + (-x)$$
$$21 + x = 15 + (-x) + x$$
$$21 + x = 15 + 0$$
$$21 + x = 15$$
$$(-21) + 21 + x = (-21) + 15$$
$$0 + x = -6$$
$$x = -6$$

Tuesday

1. $x - 11 = 19$
2. $-3k + 4k - 6 = 12$
3. $37 = 15 + c$

$$x - 11 = 19$$
$$x + (-11) = 19$$
$$x + (-11) + 11 = 19 + 11$$
$$x + 0 = 30$$
$$x = 30$$

$$-3k + 4k - 6 = 12$$
$$k + (-6) = 12$$
$$k + (-6) + 6 = 12 + 6$$
$$k + 0 = 18$$
$$k = 18$$

$$37 = 15 + c$$
$$-15 + 37 = -15 + 15 + c$$
$$22 = 0 + c$$
$$22 = c$$

$$7l + 92 = 4l + (2l + 293)$$

Wednesday

1. $22 - a = 19$

2. $18m - 44 + 79 - 17m = 32$

3. $(4l - 12) - (3l + 6) = 97$

$22 - a = 19$
$22 + (-a) = 19$
$22 + (-a) + a = 19 + a$
$22 + 0 = 19 + a$
$-19 + 22 = -19 + 19 + a$
$3 = 0 + a$
$3 = a$

$18m - 44 + 79 - 17m = 32$
$18m + 35 + (-17m) = 32$
$18m + (-17m) + 35 = 32$
$m + 35 = 32$
$m + 35 + (-35) = 32 + (-35)$
$m + 0 = -3$
$m = -3$

$(4l - 12) - (3l + 6) = 97$
$4l - 12 - 3l - 6 = 97$
$4l + (-12) + (-3l) + (-6) = 97$
$4l + (-3l) + (-12) + (-6) = 97$
$l + (-18) = 97$
$l + (-18) + 18 = 97 + 18$
$l + 0 = 115$
$l = 115$

Thursday

1. $9 + y - 4 = 2y - 6$

2. $(4x + 6) - (2x - 9) = x + 14$

3. $k + k + k + 7 = k + k + 6$

$9 + y - 4 = 2y - 6$
$9 + y + (-4) = 2y + (-6)$
$9 + (-4) + y = 2y + (-6)$
$5 + y = 2y + (-6)$
$5 + y + (-y) = 2y + (-y) + (-6)$
$5 + 0 = y + (-6)$
$5 + 6 = y + (-6) + 6$
$11 = y + 0$
$y = 11$

$k + k + k + 7 = k + k + 6$
$3k + 7 = 2k + 6$
$-2k + 3k + 7 = -2k + 2k + 6$
$k + 7 = 0 + 6$
$k + 7 + (-7) = 6 + (-7)$
$k + 0 = -1$
$k = -1$

$(4x + 6) - (2x - 9) = x + 14$
$4x + 6 - 2x + 9 = x + 14$
$4x + (-2x) + 6 + 9 = x + 14$
$2x + 15 = x + 14$
$-x + 2x + 15 = -x + x + 14$
$x + 15 = 0 + 14$
$x + 15 + (-15) = 14 + (-15)$
$x + 0 = -1$
$x = -1$

$11y - 22 - 33 = 10y + 55 + 66 + 26$

Friday

1. $14z + 18 = 12z + (6 + z)$

2. $12 - 6x - (3 - 5x) = 30$

3. $6b + 14 - 4b = 44$

Wednesday's third hunt was a mess. Marco had erased and restarted so many times the spirts of his mistakes still haunted the page. It was the parentheses that caused the problem.

At first, he had changed $(4L - 12) - (3L + 6) = 97$ into $4L + (-12) + (-3L) + 6 = 97$. He sent in duel after duel to reveal 103 was hiding behind L. The problem was, when he tried to check it, he ended up with $400 - 315 = 97$ and finally $85 = 97$ which was a clear indication to Marco he had made a mistake. Almost feeling guilty for jailing the wrong Numberfolk, he called on his sister again to help. After a very long winded and twisty explanation, Marco understood that the whole thing in parentheses was being subtracted, not just the $3L$.

After promptly forgetting Maggie's reasoning around subtraction of parentheses, he instead imagined an old western saloon. Inside the bar sat four L's and a -12. Suddenly, a band of robbers burst through the swinging doors. "This is a stick up, nobody move!" They shouted. Demanding a ransom of three L's and (hence the plus) a 6, the owner had no option but to offer up three of the four L's in the bar.

"Go ahead and take 'em!" he shouted to the bandits. But he still owed them a 6. "Why don't ya take six from thar -12?"

The -12 stood, ready for a fight. The robbers pulled out a -6 from their wagon and sent them over to -12 to get what they demanded. The two numbers floated their hands by their sides, ready to duel. The clock bell rung, and they were off, tables and chairs flying around as they scuffled. When the dust cleared all

$$5y + 68 = 4y + 271$$

that stood in the bar was the one lone L the robbers had left behind and a -18, the result of the duel between -12 and -6.[°]

Marco had used the same tactic on his Thursday question. He looked over his work with satisfaction. In each hunt, he sent in duel after duel to eventually isolate and reveal the numbermask. As he examined his battles, he took on the role of a military commander. "Send in a six! Hold firm, hold firm! We almost have the enemy right where we want them!"

Scanning the rest of the page, he thought about how happy Mr. Pikake would be when he saw all his hard work. *Alright. Three more to go.* Marco prepared himself for battle.

The first hunt was a familiar foe. Numbermasks stood everywhere – he knew this would require a few duels to find z, but he also knew he could do it.

$$14z + 18 = 12z + (6 + z).$$

He combined the like terms on the right. He had $12z$'s plus another which left him with $13z$'s and the 6.

$$14z + 18 = 13z + 6.$$

Then, Marco reached down as if grabbing something from beneath his desk and threw it at the paper. He had kidnapped 6's evil twin and sent it to battle!

$$14z + 18 + (-6) = 13z + 6 + (-6).$$

When the dust cleared, the 18 had dueled with -6, leaving 12 in their place and the evil twins vanquished each other, leaving Zero to battle with the $13z$'s.

[°] Marco found that $(4l - 12) - (3l + 6)$ was *not* the same as $4l - 12 - 3l + 6$ as he thought. Since the entire parentheses are being subtracted, it is the same as distributing the negative. So $(4l - 12) - (3l + 6) = 4l - 12 - 3l - 6$. Both terms in the parentheses must be subtracted or as Marco sees it, taken by those pesky bandits.

$$-1 - 2 - 3 - k = -k + 65 + 66 + 67 - k$$

$$14z + 12 = 13z + 0.$$

Marco reached below again, this time throwing a -12 at both sides.

$$14z + 12 + (-12) = 13z + (-12).$$

He blinked. And then again. He had $14z + 0 = 13z + (-12)$. He wasn't sure that last move helped the situation any. Marco wanted to erase the last step, but this wasn't a Marvel movie. He couldn't go back in time and change the past – this was war! He decided to push forward.

$$14z = 13z + (-12).$$

Fight fire with fire, fight masks with masks, he thought before throwing $-13z$ into battle.

$$14z + (-13z) = 13z + (-13z) + (-12).$$

On the left, his $14z$'s where diminished to only one when they dueled the $-13z$. On the right, well, a grin spread across Marco's face when he saw the result.

$$z = 0 + (-12)$$
$$z = -12.$$

Marco imitated the crowd roaring at his victory as he moved forward to the next round.

$$12 - 6x - (3 - 5x) = 30.$$

Ah, this battle would take place in the saloon. A 12 and six $-x$'s sat around the tables. The robbers entered as expected. Their demand? "Give us 3 and five $-x$'s!" one of them shouted. The whole situation of subtraction with parentheses still confused Marco. As a result, the $-x$'s were confused, too. Do we go left? No right? Left. They shuffled back and forth.

The bar owner slammed his hands on the counter. "They demanded five $-x$'s!" he shrieked. "There be six ov' you right thar." He pointed to the door and five of the six $-x$'s slowly

$$x + x - 25 + 75 - x = 255$$

marched out. The one lonely $-x$ sat shaking in the corner. Marco looked around to assess the situation. A 12 sat at the counter, the bandits were yelling, demanding their 3. Marco obliged. He reached into 12 and pulled out 3, which he threw to the robbers. The 12 transformed to a 9 and angrily glared at Marco. All in all, they survived the heist. Only the 9 and the one $-x$ were left in the bar, but the robbers had gone away, happy with their take.

$$12 - 6x - (3 - 5x) = 30$$
$$12 + (-6x) + (-3) + 5x = 30$$
$$12 + (-3) + (-6x) + 5x = 30$$
$$9 - x = 30.$$

From here, it was easy for him now. He made quick work of the remaining steps. He sent in x to duel which gave him:

$$9 + (-x) + x = 30 + x$$
$$9 + 0 = 30 + x.$$

Then he sent in -30 to wrap things up.

$$-30 + 9 = -30 + 30 + x$$
$$-21 = 0 + x$$
$$-21 = x.$$

He held up his pointer finger and blew at it as if extinguishing a candle. Last one.

$$6b + 14 - 4b = 44.$$

Rewriting as a duel and rearranging first, Marco wrote:

$$6b + (-4b) + 14 = 44.$$

He combined the like terms. This time, there were six vampires talking merrily. One of them brought up politics which was a major faux pas causing four to kindly exit the conversation. The remaining two discussed the blood banks cheerily as Marco was left with:

$$2b + 14 = 44.$$

$$6r - 27 = 7r + (87 - 2r) + 92$$

Panic instantly set in. This was a situation he hadn't seen before. In all the hunts, he was always left with *only one mask*. Here he had two: there were two b's hiding together. His hand shook a bit as he followed his algorithm, sending in 14's evil twin to duel.

$$2b + 14 + (-14) = 44 + (-14).$$

Wearily, he continued.

$$2b + 0 = 40$$
$$2b = 40.$$

"Crap." He went through his previous moves trying to find his mistake, but his hunting was solid. Everything he had done made sense, but his answer, unfortunately, did not. He hadn't got the numbermask alone, he couldn't complete his hunt. Marco stared at the paper. And stared and stared. He welcomed the blurry characters that came when his eyes could no longer focus, and he didn't even try to readjust. He stared, without blinking, until tears formed in his bottom lids. He stared until his head started hurting. He stared until his helplessness turned to anger.

Throwing his hands across the desk, pencils flew around the room. He was a failure. He was 'not a math kid'. Who was he kidding? Why did he ever think he could do this? Because he's stupid and that's what stupid people do – believe in the impossible. His thoughts were his own but the voice yelling in his head was Peter's. Filled with an uncountable number of emotions, Marco pushed himself away from his desk and went to the bathroom. The sweat from baseball practice had dried, stuck to his skin, and now nervous sweat was dripping on top of it mixing with the old sweat to form a new sticky substance. He turned on the shower.

Letting the running water wash over him for what felt like a long time, Marco tried to force his mind to go blank. He didn't want to think about math or Numberfolk or hunting, he didn't want to think about baseball or his friends or his sister. He didn't want to think about his mom or Peter or Mr. Pikake. When Marco finally opened his eyes, he was relieved to find he was starting to

$$-3y + 45 = -(y + 73) - (y + 89)$$

feel better, lighter. In front of him, a ledge held Maggie's surplus of beauty products. *How can one person need this much junk in the shower?*

Trying to keep his mind on anything other than life, he began reading the bottles of her shampoo and conditioner. As he scanned to the very bottom, he saw it. Each bottle contained 20 fluid ounces, which meant together they must be 40 fluid ounces.

In his very own *Eureka!* moment, Marco finished and jumped out. He looked back at the 20 on the bottles and said, "This is your last day. Tape is coming your way tomorrow!" Although the numbers had been helpful, he certainly didn't need them everywhere watching his every move – especially in the shower. He made himself a mental reminder to tape over all the numbers in the bathroom before throwing on his pajamas and bouncing back into his room.

He stared down the equation $2b = 40$, this time as its master. The equation was no longer in charge. Marco was in charge. He let the powerful feeling overtake him before very slowly and very neatly writing below it: $b = 20$. The whole thing now made perfect sense. But his process wasn't strong enough to tackle this type of situation again. He couldn't replicate it or hope Maggie's bottles would come to his rescue next time. He wondered if there was an algorithm for these types of hunts, for when the Numberfolk weren't using duels to hide but instead under enlargement spells making them stronger, more conniving.

After saying goodnight to his mother and sister, he didn't even know where Peter was, Marco laid in bed staring at the ceiling. He tossed and caught a baseball until his eyelids were too heavy to hold open. Rolling to his side, Marco fell asleep.

All night long Marco had dream after dream, all the same but slightly different. They always started with him walking through the streets of London. Well, he thought it was London, he had never actually been there, but it was what he imagined the city looked like. Out of nowhere a zombie latched onto his arm, its

$$(14m - 32) - (6m + 14) - (7m + 33) = 129$$

teeth gnawing for a bite of tasty flesh. Whacking it with a bat that had appeared in his hand (or was he always carrying it?), he hit it again and again until the zombie was nothing more than a pile of goop at his feet.

He continued forward down the street battling zombie after zombie, easily defeating any undead obstacles that stood in his way. When he reached the end of the road, a giant structure lay in front of him. He craned his neck back to examine it. The building was white, too white, a white that was so clean and pristine it could never exist in the real world. It was covered with intricate carvings, and four large spires shot out high from each corner. Unsure where he was, Marco decided the building must be one of those old gothic churches he had seen on TV. Four stairs led to an enormous arched entryway complete with dark, wooden doors. Ascending the stairs, Marco pushed hard on the doors, they were unnaturally heavy. It took everything he had to create a crack large enough for him to slip through.

The doors slammed shut as soon as Marco released them, almost smashing him in the mouth. In front of him stood a long aisle with benches lining each side. At the very end, what looked to be almost a mile ahead, a giant zombie lay chained to a pool at the altar. Upon sight of Marco, the beast began thrashing and pulling at its chains. Looking down, Marco's hands were empty. What happened to his bat? His memory was fuzzy, like the bat had never really existed. It didn't matter, it wouldn't do much good against the giant. As the zombie pulled and pulled, trying to reach Marco, it finally ripped its chains. Now able to unfold, the monster stood at its full height. Its head landed between the thick wooden rafters that sat at the top of the room over thirty feet above the ground. Marco ran.

The entryway stretched farther and farther away. Knowing he would never get the heavy doors open in time, he changed strategies and ran straight *towards* the beast. The giant swatted at Marco but was too slow. He was safely, at least for now, between the tree-trunk legs the beast stood on. Marco grabbed a wooden

$$x + 69 + 70 = (x + 1) - 71 + x$$

railing and with a sharp downward force broke it in half to reveal a large stake. He jabbed the stake into the giant's leg. It howled in pain, bending over and swatting with both arms determined to end what to it must have seemed like a pesky, yet dangerous, fly.

Marco zigged and zagged between the giant zombie's legs until he found his opportunity. With another sharp blow, he stabbed his stake into its unharmed leg. The zombie wailed and wobbled before falling flat on his face. The entire building shook from the impact. Not wasting any time, Marco ran up the monster's back and plunged the stake once more. The giant disappeared beneath him.

Eager to escape, Marco sprinted to the entrance. With all his strength, he threw himself into the doors which flew open – as if suddenly they weighed almost nothing at all. As he stumbled outside, the street was no longer desolate. Crowds of people had gathered on each side. They were all chanting Marco's name, cheering, waving their hands in the air. Marco pumped his fist up high. He soaked in the adoration, feeling accomplished: a warrior.

In other versions of his dream, different terrifying enemies waited for him behind the church doors. Once was a pack of rabid wolves, another time it was a horde of zombies, and one time it was the mayor from Marco's video game. No matter what presented itself, Marco was victorious. It was never easy, but he always made it back outside to the crowd of admirers chanting and cheering in front of him.

It was then that the dream turned to a nightmare. In every variation Marco would begin down the street, through the crowd. A large, black hawk jumped from its perch atop the left spire and soared down. As it approached, Marco could see its bloodshot eyes and thick drool dangling from its beak.

Once, he tried to run. Another time he tried to fight. He tried to hide and throw himself into the crowd to vanish in the sea of people. No matter what he did, the bird eventually caught Marco and tore at his flesh until he woke up, heart pounding and

$$x + 105 = 2x - 105$$

drenched in sweat. Once, he caught the hawk's eye just right to see the flames of a hot fire glimmering within. Marco could swear he recognized it – but couldn't remember where he had seen it before.

$$21 + y + 42 + y + 2 = 91 + y + 92 + 93$$

13

THE EXPANSION OF NUMBERVILLE

"WHEREVER THERE IS NUMBER, THERE IS BEAUTY."

-PROCLUS

F or many years after the Wars, the state of Integer was at peace. As Numberfolk settled down, they did as most beings do – they began to form families.

The first numberbabe was born to a five and a four. No one had any idea what would happen when two Numberfolk produced a child, so all eyes were on the couple. When the baby finally arrived, the village of Natural was buzzing with conversation. Most believed the child would also be a four or a five. Their logic was that the baby would be a hybrid – a mix of its parents. They reasoned that since half of four was two and half of five was between two and three, it could have swung either way. Gaining two from four, the child could gain another two from five or three from five. No matter what, they predicted the offspring would be the image of one of its parents. As googols* of eyes peaked into the nursery to look upon the very first numberbabe, they were both shocked and delighted to see a tiny twenty, giggling with life.

Enter the Thinkers. All of Integer was eager to understand what happened when Numberfolk multiplied. As their society was relatively young, the Thinkers always had something new to explore. After much debate, the group felt they understood reproduction and presented their beliefs on the multiplication of Numberfolk.

> *When Numberfolk produce a child, there are two possible events that may happen. What we have witnessed in our examinations is a series of duels between the genes occurring.*
>
> *Suppose a is one parent and b the other. The child of a and b can be either be the result of a duels with b or the*

---

* A googol is 10,000,000,000,000,000,000,000,000,000,000,000,000,000, 000,000,000,000,000,000,000,000,000,000,000,000,000,000,000,000, 000.

result of b duels with a. It so happens that despite which series of duels occurs, the result is the same.

For example, we posit that should a six and a ten have a child, the child would be born as the result of six duels of tens (that is $10 + 10 + 10 + 10 + 10 + 10$) or the result of ten duels of sixes (that is $6 + 6 + 6 + 6 + 6 + 6 + 6 + 6 + 6 + 6$). Regardless of the gene battles that occur, the product of a six and a ten, would therefore be a sixty.

Furthermore, the genetic process includes two steps. First, the genetic material from one parent directs the child to create clusters. Next, the clusters are fused together through dueling – the number of duels determined by the second parent. In the case of a six and a ten, step one would be the creation of clusters of size six, and step two the fusion through ten duels of the clusters. Or alternatively, the creation of clusters of size ten followed by the fusion through six duels of the clusters.

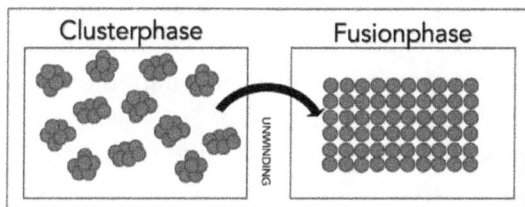

These findings were amazing news. For in the history of the world, long after this story occurred and beyond, it has been the pattern of almost every civilization that the child of two parents always is born into a similar societal class. This was not the case in Numberville. For any two Naturals who multiplied would find their resulting child to be a larger value, to be of greater magnitude, than its parents.

After this discovery, many residents in Natural began to multiply and the entire village welcomed the new residents. They specifically loved that every numberbabe was again a counting number – another

Natural, their own kind*. They were one big happy family all living together within their village. That is, of course, until everything went astray.

It wasn't long before a Natural and a Negative fell in love. And why shouldn't they? The story that has been passed down is that the first to multiply was a one hundred and a negative two. At this time, all of Integer exploded with theories.

"It will clearly be one hundred negative two's," some residents claimed.

"I imagine it would be two negative one hundreds," others chimed in.

Some laughed thinking the child would be a freak of nature – negative two one hundreds – although no one at all had any idea what that would even look like.

The day came and the baby was born. It was a chunky little -200 that was as adorable as you could imagine. The baby was ripped from One hundred's arms and carted off to TNZ. One hundred was devastated. While he was able to visit and care for the child, it was impossible for his baby to live with him in Natural, all Negatives were required to live in TNZ.

After this, it was rare for a Natural and a Negative to have a child, for no Natural wanted to live apart from their baby. This did however pave the way for a program that allowed residents of TNZ who wanted a child, but did not have a partner, to multiply with a Natural and take full responsibility of their kin in TNZ.

Despite the fact that everyone knew the product of a resident of Natural and a resident of TNZ would be a Negative, the Thinkers again published their take on the situation.

It has come to our attention that some Numberfolk have begin to doubt our publications due to the new discoveries

---

* This is called closure. Integers are closed under multiplication.

regarding the multiplication of residents from across Integer. While we stand strong with our original conclusions, we attempt to provide additional clarity.

When two residents multiply, their product is a mixture of the genetic material of both parents. As previously stated, the creation of a numberbabe occurs in two steps: the generation of clusters and the fusion of those clusters through dueling. We now also recognize that either step may occur in what we are calling the 'right-side up' or the 'upside-down'.

We have long known that residents of Natural are composed of positive particles while residents of TNZ are composed of negative particles. It is these particles that impact the child's development.

Consider the product of parents a and $-b$. In the first step of the process, the babe may construct clusters of size a in the right-side up. The second step tells the child to fuse b clusters together. However, because b is a Negative, its particles require construction in the upside-down and the ultimate result is a child of TNZ.

If $-a$ and b were to produce a child, the clusters of size a would form in the upside-down as a is a Negative, then b clusters are fused together and remain in the upside-down, again producing a resident of TNZ.

Some concern has arisen about the possibility of numberbabes not being Integers at all – innocent children sent to live unaccompanied in the barren, empty land outside the state. Rest assured, we have confirmed that any product of residents of Integer will again be an Integer and guaranteed housing within the state.

We conclude our publication with some examples to better clarify the genetic multiplication process.

Consider the numberbabe produced from a two and a five. As expected, the genetic material emits a code instructing the babe to produce clusters of size two. As two is a Natural, with positive particles, the clusters are built in the right-side up. Next, the babe is told to fuse together five of the two-clusters. Again, as five is a Natural, this process too occurs in the right-side up. The result: a positively charged numberbabe ten.[*]

Now consider the numberbabe produced from a two and a **negative five**. The first step is the same. The genetic material instructs the babe to build two-clusters in the right-side up. However, in the second

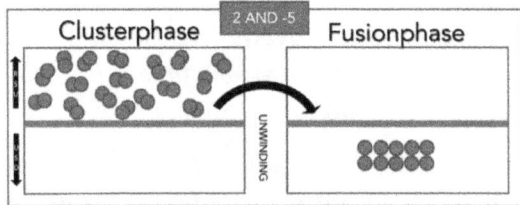

step, the **negative five's** particles instruct the babe to fuse together five of the two-clusters in the upside down! The negative charge ultimately tells the genes to build in the opposite direction. Since the clusters were in the right-side up, the introduction of the negative materials directs the clusters to form in the opposite space – the upside-down. The result: a numberbabe ten being made of five clusters of two, but a negatively charged ten, a ten built beneath, in the upside-down. A negative ten[†].

What would be produced by a **negative two** and a five? Well, in this case, the genetic material begins to build clusters of size two in the upside-down. In step two, five of the two-clusters are fused in the same direction. The result: a numberbabe of size ten but being constructed in the

---

[*] We denote the numberbabe of a two and a five today as $2 \cdot 5 = 10$.

[†] The Thinkers' report in modern day notation tells us $2 \cdot -5 = -10$.

upside-down the babe has a negative charge and thus results in a negative ten[*].

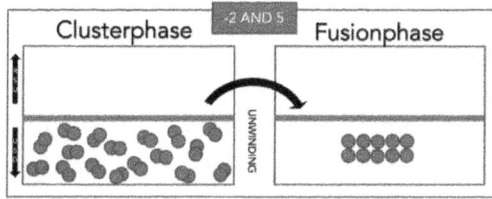

We hope this publication provides the clarity the residents of Integer demand from their Thinkers.

The Thinkers were right. Many residents were losing confidence in the group as they had failed to predict the product of a Natural and a Negative. However, this new publication helped restore faith across the land. That is, until the first baby of TNZ was born. It is not known why residents of TNZ waited to begin families, but it was almost a year later before their first was born.

A negative three and a negative five were the inaugural pair to multiply and the result was devastating. Excited to start their family, when they welcomed their child into the world, they were permitted only a glimpse before it was abducted from their home and whisked off to Natural. The residents of TNZ were outraged, and it seemed like another war was on the horizon. There was a much different feeling in Natural. The overwhelming majority felt like the result was just, it was only fair. They believed this twist of fate was how their ancestors maintained balance in the world. Just as the Great Scale had created TNZ, she also had the power to take away. These feelings were likely the result of lingering anger over Natural/Negative babes being automatically sent to TNZ.

In no time at all, the Thinkers produced their third and final pamphlet on the subject.

Residents of Integer, we come to you to discuss the matter of the product of two residents of TNZ in our writings on the process of multiplication.

---

[*] Similarly, $-2 \cdot 5 = -10$.

We have previously posited that the product, the child, of any two Naturals is again a Natural. Further, the product of any Natural and any Negative will ultimately be a resident of TNZ. In describing the process once again, we hope to quell any outrage around the numberbabe of two Negatives.

Recall a numberbabe is first instructed to build clusters, its parents being both Negatives ensures that these clusters will be constructed in the upside-down. Next, the babe fuses the clusters together. As previously posited, should parent two be a Negative, the clusters are instructed to turn around – to build in the opposite direction. Since the clusters were formed in the upside-down, this tells the babe to fuse their clusters in the right-side up and unsurprisingly, the offspring is a Natural.

It has been well documented that TNZ is in many ways the opposite of Natural. This has been observed in the business of duels, as well as in the basic structure of TNZ overall. When Naturals duel, the result is a larger Integer. Yet, when Negatives duel, the result is a smaller Integer. So, it comes as no surprise to the Thinkers that genetic duels have analogous behavior.

We include an example to complete our chronology.

*Should a **negative six** and a **negative five** choose to multiply, in the first step, the numberbabe is instructed to form clusters of size six in the upside-down due to the negative particles of the six. In step two of the process, the babe is told to fuse together five of these clusters. Due to the negative charge of the five, the babe is directed to fuse these clusters in the opposite direction. Because the clusters were built in the upside-down, the babe*

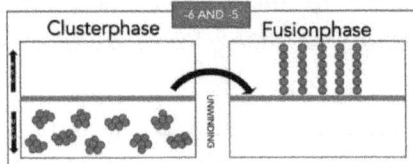

at no fault of its own, completes the fusion in the right-side up. The result: five clusters of six in the right-side up. The numberbabe shall be a Natural.*

Now, there has been much ado about fairness and equality. It is our conclusion that the multiplication outcomes are exactly impartial. Consider this: there are four options in which two Integers may multiply. Parent 1 may be either a Natural or a Negative and Parent 2 may be either a Natural or a Negative.

First, consider the cases in which Parent 1 is a Natural.

- *If Parent 2 is a Natural, the numberbabe is Natural.*

- *If Parent 2 is a Negative, the numberbabe is a Negative.*

Next, consider the cases in which Parent 1 is a Negative.

- *If Parent 2 is a Natural, the numberbabe is Negative.*

- *If Parent 2 is a Negative, the numberbabe is Natural.*

Here we can see that exactly half the outcomes are Naturals and half the outcomes are Negatives proving the equality of reproduction. Further, we again can witness the opposite power residents of TNZ possess. For when parent one is a negative, the product is precisely the opposite of parent two.

	Natural (N)	Negative (Ng)
Natural (N)	N	Ng
Negative (Ng)	Ng	N

Table 1: Description of the product of Integers

---

* Sadly for TNZ, this was to say $-6 \cdot -5 = 30$.

The final pamphlet from the Thinkers[*] was successful in calming the outrage just enough to avoid war. Many accepted that multiplication was fair and the products of Numberfolk were evenly split between Natural and TNZ. However, parents on both sides longed for their children who were forced to live in a different land. Luckily, numberbabes were always Integers and could easily spend time together around the state.

This concludes our overview on the expansion of Numberville. We will leave you with the *Fable of Seventeen*.

THE FABLE OF SEVENTEEN

Seventeen was a unique Numberfolk. For she loved herself far more than she could ever love another. Many believed this narcissism was a result of her being prime. She felt no connection to others as she could not be built with twos, or threes, or fours, or fives, or any other combination. Seventeen was made up of only herself and this showed in everything she did.

Her vanity was first noted in the time of the duels. Refusing to duel with anyone but zeros, Seventeen simply adored the result of her battles with the sons of Zil. She would duel again and again with zeros, each time seeing only her own reflection. As other Numberfolk began to multiply and produce families, Seventeen became bored and lonely. She wished to have a numberbabe of her own.

It seemed only right that she would produce a child with a Zero. She loved Zero, if only for the fact Zero always showed her exactly

---

[*] Readers may be interested to know that human genetics work astonishingly similarly to Numberfolk genetics. The table presented to Integer from the Thinkers bears shocking similarities to the Punnett Squares used to describe dominant and recessive genes in humans.

what she wanted to see – herself. It wasn't long before the two welcomed a numberbabe. While the child was no surprise to fellow Numberfolk who had followed the Thinkers' pamphlets closely and knew the genes would either form seventeen clusters of zero or zero clusters of seventeen, Seventeen herself didn't bother with such things that were not focused on her. She was disgusted when she first laid eyes on her child – a healthy, handsome, and charming little zero. Furious at her partner, Seventeen yelled and screamed, "How dare you produce your twin and not mine!" before abandoning the baby in Whole and returning to Natural.

Seventeen spent many years moping through the village crying. She looked on in jealousy at the loving families and their numberbabes and wondered why she could not have what they did. Why was she cursed to be alone?

One day, as Seventeen was pouting, dragging her feet around the village feeling overwhelmingly sorry for herself, she noticed something amazing. Peeking into the window of one of her neighbors' homes, she saw a quaint family. It was a one and a five and their three young children. Normally, the sight of a family made Seventeen sick with rage – but not this time. For as she peered into their home, she noticed that every child was the spitting image of their parent – every single child was a five!

She raced home and began studying the publications on producing a child. Maybe, just maybe, all her dreams could come true! Deciding that if she multiplied with a one the genes would either form seventeen clusters of one or one cluster of seventeen, she threw the papers in the air and danced around merrily.

It was not long before Seventeen found a lonely one and desperately courted him. The couple went on to multiply and produce an enormous family – each child the spitting image of their mother. As Seventeen sat in their home, looking upon her children and seeing herself in each one, she was finally content.

$3^2 + 5$

$(2\sqrt{7}\cos 45°)^2$

2_6

$41 - x = 3^3$

$\dfrac{182}{13}$

$\sqrt{196}$

$1^2 + 2^2 + 3^2$

14

MULTIPLICATION

*"WITHOUT MATHEMATICS,
THERE'S NOTHING YOU CAN DO.
EVERYTHING AROUND YOU IS
MATHEMATICS. EVERYTHING
AROUND YOU IS NUMBERS."*

-SHAKUNTALA DEVI

A strong punch of sunlight hit Marco in the face. He groaned and rolled over. As he blindly groped his nightstand sending things flying to the floor, he finally found what he was searching for. Prying one eye open, he quickly glanced at his phone. *Nine hours?* While the clock was telling him he'd slept all night, his body was saying something very different. The series of nightmares had left him exhausted.

He forced himself onto his feet and, still in his pajamas, drug himself down to the office. Liam and Oliver were already online waiting for him. Things had started to cool with Oliver. While their friendship wasn't back to normal, Marco didn't feel like he was constantly under attack either. The team had made it over the mountain with some stellar new gear and had even picked up a few NPCs from the cave to tag along and provide extra protection. As they hiked down the road in search of their next mission, far off in the distance Marco could see the outline of a crow sitting atop a speed limit sign. Ignoring his surroundings, Marco focused on the bird – he couldn't take his eyes off it. Suddenly, a horde of zombies began pushing their way from the neighboring forest onto the road. As Liam and Oliver began to attack and defend the group, Marco remained frozen on the bird.

It wasn't long before the screen went red. "Marc! What are you doing?!" Oliver screamed angrily into the headset.

Snapping out of his daze Marco responded, "Sorry. I had this horrible dream last night about birds. I can't get them off my mind."

"Birds don't exist," Liam chimed in. "It's a conspiracy, you should be afraid."

Something about Liam's words pulled Marco back in time dislodging a memory. The boys were in third grade – or was it fourth? Everyone in town had been on edge because a local kid had gone missing. Liam insisted the police knew who the kidnapper was but weren't able to apprehend them. He called him 'The Algebraist' and described the man as some sort of genius – a rocket scientist. Liam went into great detail claiming the man took kids as trophies. But not any kids: smart kids. The boys joked about how at least *they* were safe. The memory started to fade, and Marco grasped on as tight as he could. What had happened in the case? Unable to remember if the boy had been found, he allowed it to float into darkness and watched as it faded away.

Giving his friends an excuse about being tired, Marco logged off. Lugging himself back up the stairs, he threw his body onto his bed. It wasn't until the familiar scream 'Wake Up! Wake Up! Wake Up!' screeched from his alarm that he realized he had slept all morning. It was worth it, he felt much better.

He made his way to the library. The weather was getting warmer, the sun's rays felt more powerful although the snow that still hugged the ground was a reminder winter remained in session. Ignoring the people bustling around him, Marco headed directly to the study room they'd reserved every weekend afternoon for the month. Immediately throwing himself into the chair he slouched down pulling his hoodie over his eyes and began to doze off. Before his first dream could take shape, the door flew open. "Good Afternooo-ooon!" Mr. Pikake boomed.

Marco, feeling like he was just caught doing something he shouldn't, bounced up, tore the hood off his head, and flashed the professor a smile. "Sorry, long night." He reached into his backpack and presented Mr. Pikake with his homework. Beaming so brightly the professor probably needed to shield his eyes, Marco shared, "I practiced my hunting every day. And I think I did a really good job. A couple things confused me but…" Marco realized he sounded amazingly like his sister and shuddered at the thought.

"But!" Mr. Pikake popped, "You preserved, persisted, and prevailed! Bravo!" He snatched the paper from Marco's hands and began reviewing his work. "Very nice! Very nice indeed!" His signature grin slowly spread from one ear to the next.

"I had a few questions." Marco looked to the professor who pushed open his eyes wide and bobbed his head forward to say 'proceed'. "I could use some help understanding what happens when we subtract a negative. And also subtracting with parentheses, it confuses me every time."

"*Ahh*, subtraction is a suspicious beast – it is." Mr. Pikake began pacing the length of the room. "I prefer never to subtract. Numberfolk subtraction was the backstabbing of the Integer Wars which was quickly prohibited. We can present any subtraction as a duel as they have done. It is helpful to look at our problems in their terms. But," a pop, "rather than opine on operations, let us conduct a contemporary review."

Mr. Pikake uncapped the green marker and wrote on the board:

$$5 - 3.$$

"This is a recognizable retraction, *ay*. How would we display this as a duel?"

"Five minus three is the same as five plus negative three. It's a duel between five and a negative three." Marco confidently stated before grabbing the blue marker from the tray and scribbling beneath:

$$= 5 + (-3).$$

"Precisely! But what have we done here? What are we saying?"

Marco had no idea the answer to that question. He thought about saying 'subtracting' but sensed that wasn't what his tutor was looking for.

"Alright. Alright." Mr. Pikake continued, noticing the boy was lost, "What if I said subtraction is the same as adding the..." the last word lingered in the air, a pitch to Marco to complete the sentence.

"Adding the evil twin?" Marco guessed.

$$(43 + 62) - (-87 - 36)$$

"Yes!" The tutor's hands flew into the air. "Subtracting three from five is the same as adding three's evil twin, which is of course negative three. Subtraction is simply a duel with a Numberfolk's twin. So, let us examine the kernel of your question." He erased the expressions and replaced them with:

$$5 - (-3).$$

"My sister told me that it is just five plus three, er." Marco always seemed to get better results when he used Mr. Pikake's wording. Quickly correcting himself he added, "The same as a duel between five and three."

"And why shouldn't it be?" the tutor chirped. "Subtraction is the same as dueling with one's twin. Here you are subtracting negative three, who is negative three's twin?"

Marco started to see it, "Three! I get it. Five minus negative three is a duel between five and negative three's twin, which is a duel between five and three, which is five plus three!" He wrote his findings on the board.

$$5 - (-3) = 5 + 3.$$

Melting into the chair, Marco looked at the board. These hiccups, blockers, lingering questions always made him doubt himself. How could he ever become a master when he didn't even fully understand something as basic as subtraction? Hearing Maggie confidently explain these ideas in the same foreign language his teachers used deflated Marco. He'd never be able to do what she does. But dueling, vanquishing values, evil twins – *that* he could understand. Mr. Pikake turned math into a game world and Marco had been mastering those for as long as he could remember.

"For good measure, we should also describe the backstabbing strategy of the wars. Do you recall the story?"

Do I recall the story, Marco scoffed in his head. *Of course, I do!* A tale about warriors breaking the rules and turning on their dueling partners was always more likely to stick than a list of mathematical rules. "Yes! Backstabbing was when they would turn and duel meaning only one of the combatants was engaging. It ended up being like giving themselves away."

"Wonderful! If you prefer to use the backstabbing strategy, we must remember what occurred. In $5 - (-3)$, the five and negative three were sent to duel, but the -3 backstabbed the 5. This technique removes particles meaning -3 particles are removed from 5."

"But that doesn't make any sense. How do you remove negative pieces?" Marco interjected.

"*Ah*...that is the question! The most obvious example is to consider debt. Do you know what debt is?"

Unfortunately, Marco was all too familiar with this concept. Money was often tight around their household, and he had heard his mother and Peter arguing over 'debt' before. "Yeah." He looked down and kicked at the ground. "It's when you owe someone money."

"That it is." The professor treaded carefully. "A negative can be thought of as a debt, something that is owed. If you have no money and owe me \$2, you have $-\$2$. If you do some work and make \$5, after paying me what you owe me, you would have $5 - 2 = \$3$ remaining. But what happens when you remove a debt?" He paused for dramatic effect. "You gain money! You have -2 and I subtract your debt, forgive it, you have $-2 - (-2)$ and you are back at zero, nothing owed. In the case of $5 - (-3)$, when -3 backstabbed the 5 it removed a debt of 3, and thus..."

"Thus, it added to it!" Marco jumped in. He saw it now, both ways. He felt far more excited than he should about math, for good reason. He felt like he had been chained down to the rules. Wanting to take the list he'd memorized and banish them forever, feverish enjoyment burned in his eyes. He now understood *why* the rules were what they were meaning he didn't need to memorize anything.

$$-33 - (-98) - (-110 - 55)$$

"So, subtracting a negative can be looked at as either first, since subtraction is just a duel with the evil twin and the evil twin of a negative is a positive, when you subtract a negative you are just adding a positive. Or second, you can think of it like backstabbing, removing. When you remove a negative you are cancelling out a debt, which is adding to what you have!"

To top things off, Marco allowed his imagination to take over. He was back in the saloon. A man dressed in all black stood at the counter, towering over the owner. It was the tax collector. "You still owe me three from last week," he said, tipping back his hat to look the owner dead in the eyes.

The owner shook, nervous. "O-o-f course, s-s-sir. B-b-but, I'm so sorry, I don't have it." The tax man looked around the room, his eyes stopped on the sign that read 'Rooms to Rent: 10 Gold Coins'.

"I be needing a place to lay my head tonight. I'll make you a deal. I'll give you five coins *and* subtract your debt. We'll call it even." The man agreed, and the tax collector slid the five coins across the tabletop.

The owner chimed in, "I-I-I'm sorry sir, you owe me three more coins. Since $5 - (-3)$ is $5 + 3$ when you subtract my debt and add that to the five for the room, you're left owing me eight." The tax collector thought on it for a moment and agreed that $5 - (-3)$ was eight so he slid three more coins towards the man before climbing the stairs to retire for the night.

Once the tax collector was out of sight, the barkeep ran over to his boss. "Why'd you make that deal? He'll be back for more tomorrow. You should've charged him the full room price!"

The owner smiled, "We came out on top on that one." He looked around to make sure no one was listening before explaining. "Say I charged him the full 10, I'd still owe him the 3. Once I paid him what was due, I'd have ended up with only 7 coins." He opened his hand to show the eight coins to the barkeep.

$$75 - (-76) - (-80)$$

Shocked the barkeep responded, "How'd you manage that?"

The owner chuckled, "E'rybody knows the tax man ain't no good at math. He said he'd give me five and subtract my debt which is giving me eight. E'rybody knows that." He let out a loud laugh.

"He thinks he got a deal – only paid eight for something worth ten, but we be the winners here." And with a wink he slipped the eight coins into his pocket.

Trying not to smile, Marco focused back on the professor who continued, "Now, parentheses are more tricky. Remember, parentheses are a Numberfolk home. Another hiding place and another way for them to protect themselves. Subtracting a full house is a duel with the home's evil twin. Yet we often find ourselves unsure of who exactly the twin is. And that!" he slammed his palm on the table, "is by Numberfolk design. You see, they are working together to cloak and camouflage. Parentheses are a cover, helping them to disguise their true nature."

Hearing that parentheses were a Numberfolk design to increase difficulty was refreshing. It meant the fault didn't lie completely with Marco. Like a boss battle, it was acceptable to fail a few times before ultimate success. Even better, this news lit a fire inside him. He now, more than ever, wanted to gain the skills needed to eliminate these deceiving foes.

"Our next step is to learn more about multiplication. Long ago, Numberfolk went through a very dark time. It is a stain on their history. They emerged from this time smarter, better able to trick us. Before we investigate hunting when there are multiple masks, or masks that have been split into parts, let us discuss some of these smokescreen systems."

Marco nodded. He sat up and leaned in. Imagining an evil wizard who entrenched a town in dark, thick, purple smoke, Marco emerged standing strong and tall, defeating the wizard and restoring the town to its original glory.

$$(43 - (-68)) - (-98 + 30) + 53$$

"Working on a team has advantages and disadvantages. The terms within the parentheses are protected in a sense. But they are also grouped together, meaning someone can easily attack everyone inside."

Understanding completely, Marco burst out, "I get it! Like when my friends and I play this video game, we generally stick together. Strength in numbers and all. But that also means attacks make us vulnerable. When we were playing this one game, we could never win together. The raiders would attack us, and they could hit us all because we were in this tight group. We finally beat them by splitting up, so any one hit wasn't as damaging."

"Simply sensational!" Mr. Pikake's ginormous smile ate his face. He hopped up and wrote on the board:

$$4(x + 6).$$

"The four, well that's the raider's attack. Because the $x + 6$ is grouped together, it hits everything in the group." He added another expression:

$$4x + 6.$$

"Now, because the $x + 6$ isn't in a group, the four is only effecting the x. While these look similar, they are actually *very* different." He glanced at Marco trying to decide if the boy was following. Content with what he saw, Mr. Pikake asked, "Can we vanquish the six in both of these instances?"

Marco understood that when there were parentheses, the four was like a spell cast over everything inside. He also knew about the distributive property but wasn't sure what that had to do with vanquishing. He walked to the board and wrote:

$$4(x + 6 - 6).$$

"Can't I just do that to vanquish the six?"

Angrily, Mr. Pikake slashed at the board drawing a thick X over Marco's work. Marco recoiled. He had never seen his tutor act this way before.

$$(89 - 24) - (-89 - 79)$$

"No. You. May. Not!" Mr. Pikake snapped. Seeing the look of fear on Marco's face, the professor softened. "I am, I am sorry, Marco. I shouldn't have raised my voice. I just, I just want you to be safe. I simply cannot allow the Numberfolk to fool you. It is my fault. I have not prepared you satisfactorily." He sunk into the chair. Marco placed his hand on his tutor's shoulder to show forgiveness.

After a moment of silence, Mr. Pikake slowly began. "You see Marco, the four has affected everything in the parentheses. They call this a Proliferation spell. Do your games have spells?" The question must have been rhetorical because he didn't pause long enough for Marco to even consider an answer. "Remind me one day to tell you all about the Visionaries and Proliferation. For now, you only need to recognize that the four is an enlargement spell over the entire house. We see $x + 6$, because that is what *they* want us to see. But that is not who they are...they have been cloaked."

"Who are they?" Marco said softly.

"You tell me!" The tutor's buoyant personality had returned which allowed Marco's tension to begin to fade.

"I remember. We learned this before. The new stuff and the old stuff is mixing up. You use the distributive property. You have to multiply the four to everything inside the parentheses."

Mr. Pikake grabbed Marco by his shoulders and lifted him a few inches off the ground, "Wonderful, my boy!" Laughing, Marco sat back down at the table. He was feeling more at ease but didn't want to risk writing the wrong thing on the board again.

"It *looks* like a 6, but it is really a 24. The four is enlarging everyone inside the house." Mr. Pikake emphasized each word to indicate its importance. "When we have no parentheses, we can vanquish freely." He pointed to the second expression, "Who do we send in to duel here?"

$$121 - (13 - 42) - (-84)$$

"We can send in the 6's evil twin, −6 because there is no cloaking spell, there is no parentheses." Marco responded, his voice upbeat.

$$4x + 6$$
$$4x + 6 + (-6).$$

Mr. Pikake added to the board. "Now what about when there are parentheses? What can we do?"

Marco wasn't sure. He was confident he couldn't do the same thing and send in the −6, that was a mistake he would never make again after the professor's outburst. Treading carefully, he guessed, "Can we distribute the four to see who is really in the house, and then use that to vanquish?"

"Precisely! The four is disguising who is inside. When we distribute it to everyone in the home we have," he finished his sentence on the whiteboard.

$$4(x + 6) = 4x + 24.$$

"So, it is really 24!" Marco exclaimed, "We send in −24 to vanquish it."

"Perfection," Mr. Pikake purred and softly smiled. He pulled Marco's hunting practice across the table and began to study it. "I am really quite impressed, son. You have mastered the art of vanquishing. You were not scared to force multiple duels to get the numbermask alone. This is remarkable!"

A smile crept onto the student's face. He tried to hide it but couldn't help how proud he was of his work, and even more, how proud Mr. Pikake was of him.

"Will you show me how you hunt?" he gently requested. Marco smiled and nudged his head forward. He was ready to show-off his skills.

Mr. Pikake flipped the page over and wrote:

$$27k - 82 + 69 = 48 + 25k - 15 + k.$$

$$99 - (-89) - (50 - 97)$$

With kind eyes he looked to Marco, "Proceed."

"Okay." Marco cleared his throat and set up straight. "First, I would change everything to a duel to better know what I am working with."

$$27k + (-82) + 69 = 48 + 25k + (-15) + k.$$

"Next, I would go ahead and complete the duels I can...the ones I know the outcome of."

$$27k + (-13) = 48 + (-15) + 25k + k.$$

"I rearranged the terms on the right to get a better view." Marco snuck a look at his tutor. The professor seemed interested in Marco's strategies, and the slight grin told Marco he was doing okay.

$$27k + (-13) = 33 + 26k.$$

"I combined the like-terms on the right. If I have 25 k's and get one more k, I have 26 k's." Mr. Pikake nodded and widened his smile, Marco kept going. "Now, I can't do anything else because my terms are different. On both sides, I have one with a k and one without a k. I need to get the k's by themselves. So first, I send in -13's evil twin to duel."

$$27k + (-13) + 13 = 13 + 33 + 26k.$$

"May I interject?" Mr. Pikake asked in a delicate tone. "Before, when you completed the duels between -82 and 69 or between 48 and -15, you didn't do something to both sides, only one. But when you dueled -13 and 13, you sent the 13 to both sides, why?"

Marco saw what Mr. Pikake was doing. The professor obviously knew the answer to his own question, he was testing his student. Feeling the smile beginning to tiptoe back onto his face, Marco held it back by talking. He didn't want to appear too cocky, he knew just how to answer the question. Even better, he knew how to answer in the language Mr. Pikake would love.

$$89 - (-76) - 21 - (-92)$$

"The initial battle," Marco paused wanting to make sure his words were just right, "was designed to trick me. I needed to first simplify the situation. A good hunter surveys the scene, understands all the players involved and then they act." He tried hard to over-pronounce 'act' to sound like his tutor. It came out more like 'acht' as if Marco had picked up an English accent.

Now even his talking couldn't conceal the smile. "With things simplified, I could begin my hunt. Only when I am hunting, when I am initiating duels to uncover the numbermask do I have to worry about keeping the scales balanced. My previous moves didn't change anything, I just made the situation easier to see. But when I force a battle, that tips the scales. I have to force the battle to both sides."

Mr. Pikake sprung from his chair, he threw his hands in the air and began dancing a little jig. Marco finally let go and his smile took over. The two danced around for a moment before bursting into laughter and sliding back into their seats. "Finish him off, son!" Mr. Pikake cheered Marco on.

Regaining his focus, Marco said, "Okay, so I forced the duel on both sides with 13."

$$27k + (-13) + 13 = 13 + 33 + 26k$$
$$27k + 0 = 46 + 26k.$$

"Now, $27k$ duels with Zero. Zero's reflective coat means the result is $27k$. But." Marco successfully made his B pop. "The numbermasks are everywhere. I need to get them together and *alone*."

$$27k = 46 + 26k.$$

"I don't know who is hiding behind k, but I do know that whoever it is, the evil twin of $26k$ is $-26k$. So, I force a duel with $-26k$ to both sides."

$$27k + (-26k) = 46 + 26k + (-26k).$$

$$-(23 - 47) - (-89 + (-78)) + 46$$

"When $26k$ duels his evil twin, $-26k$ the result is Zero. And when $27k$ duels $-26k$, that is just saying I have twenty-seven of whatever k is and I take away twenty-six of them, so I am left with a single k. Just what I want." Marco flashed his tutor a sly look.

$$k = 46 + 0$$
$$k = 46.$$

"The last battle is the duel between 46 and 0 which, of course, results in 46. So, I did it: 46 is hiding behind the numbermask k!"

Mr. Pikake outstretched his arm to high-five Marco. With a hard slap, their hands met. "Simply superb, son." The tutor was beaming. "Before our last order of business, let us clarify your initial ponder on parentheses."

Marco had forgotten about his question and was glad his tutor remembered. He wanted to know what happened when you subtracted a negative and he felt good about that. Subtraction was just a duel between the evil twin. So, subtracting something like -5 was just adding -5's twin, 5. He still didn't quite understand the concept of subtracting parentheses. He racked his brain trying to remember everything they had talked about today. *Parentheses are like a house, a group of Numberfolk working together. When something impacts the house, it impacts everyone inside.* Mr. Pikake broke into his thoughts.

"You were successful in your similar hunts," Mr. Pikake started. "How did you manage the muddle here?" He pointed to Wednesday's third problem:

$$(4l - 12) - (3l + 6) = 97.$$

Marco excitedly explained the world he had created to his tutor. He told him of the four l's and the -12 sitting in the saloon. The bandits demanded 3 of the l's and a 6. So, three of the four l's were sent off with the bandits leaving only one l in the saloon. The robbers then took 6 from the -12 to leave -18 as well before riding off into the sunset.

$$121 - (37 - 125) - (-29)$$

Chuckling, Mr. Pikake responded, "Your imagination is nothing short of incredible." He thought for a moment before adding, "I am not sure I can do better than you have already done. You have created a marvelous way of thinking about things. I do have a key tip that is vital for every hunter to know. Remember that One has the ability to disappear!" He wrote a simple expression on the paper.

$$y$$

"How many numbermasks do you see?" He looked to Marco waiting for an immediate response. Marco hesitated. The question seemed simple, too simple. Scared it was a trick, Marco tried to uncover the catch.

Not finding one and feeling like he had already taken too long, in a loud whisper he responded, "One?"

"Yes! There is a single numbermask here. What about now?" He added below the y.

$$2y$$

"Well, now there are two numbermasks. There are two y's."

"Precisely! Our next focus will be to obliterate masquerades. That is when numbermasks disfigure themselves to hide from us. They may enlarge themselves or tear themselves into parts all as a guise to fool us."

"Like in my hunt! When $2b = 40$, I stumbled on b by pure luck. But, I don't think I could always do that."

"Exactly! To understand obliteration, you must first understand the supremacy of One. You see, while Un and Zil had the fewest Spogs, they also were given special abilities: the authority to identify Numberfolk. A good hunter uses this knowledge to their advantage. We use the power of Zil's image, Zero, in vanquishing."

"Because anyone who duels with Zero comes out unchanged!" Marco burst.

$$-47 - (-123) - (-22 - 141)$$

"*Ay*, and to get the mask alone, we need only a single mask, one. Where Zero is the identity of the duels, One is the identity of the masquerades*. And more so, One is a sorcerer who can disappear. As you deftly defined, we see y but we know that is really $1y$, an isolated mask."

"How does that help us with parentheses?" Marco wondered out loud.

"It helps because we know of the invisible one who is standing outside the parentheses house, cunningly effecting every Numberfolk inside." The professor scribbled a new expression on a blank sheet:

$$(3x + 2) - (4x - 8).$$

Before beginning, he winked at Marco. "I see this as subtracting a full wagon of Numberfolk, huddling together to survive, to hide. Since subtraction is a duel with an evil twin, my question becomes, who is the evil twin of $4x - 8$?" He looked to Marco indicating he wanted the boy to also consider the question. "Knowing there is an invisible One, I clear the battlefield to better assess the situation."

$$(3x + 2) + \big(-1(4x + (-8))\big)$$

The parentheses made Marco dizzy. It was like an infinite mirror where the shape kept repeating itself over and over.

"Now, I see that the entire house, $4x - 8$, is actually under the effect of the -1. This means it is turning everyone inside into their evil twin. The twin of $4x$ is $-4x$ and the twin of -8 is 8." The tutor wrote on the page:

---

* In mathy terms, Mr. Pikake is saying that 0 is the additive identity, anyone who goes up against him in a duel simply sees themselves. Similarly, 1 is the multiplicative identity, anyone who multiplies with 1 or is under a proliferation of 1 also emerges as themself.

$$\big(62 - (-34)\big) - (-47 + 21) + \big(87 - (-31)\big)$$

$$(3x + 2) - (4x - 8) = 3x + 2 + (-4x) + 8.$$

"Now, I have uncovered the number group! I have found who is hiding inside and can continue."

$$= 3x - 4x + 2 + 8$$
$$= -x + 10.$$

He flourished his arms like a showman before the final bow.

Making a mental note, Marco reminded himself: *When an entire house is being subtracted, something in parentheses, that is a spell cast on the house disguising everyone who is at home as their twin. To unveil who is hiding, I just find the twin of each person inside.* He snickered as he realized his two questions were the same. Subtracting a negative was just a duel with an evil twin $5 - -3 = 5 + 3$. However, subtracting a group huddled in parentheses was also the same as a duel with their evil twin! $5 - (-3 + 2) = 5 + (3) + (-2)$ since the twin of -3 was 3 and the twin of 2 was -2.

"Lastly, we must talk of inequalities. Do you know what these are?" the professor looked to Marco.

Having learned about the alligator eating the bigger number in third grade, Marco was no stranger to inequalities. Plus, these statements had the bonus of already setting up his imagination to run wild. "Yes, those are greater than or less than and stuff. The open side always points to the bigger number."

"Wonderful, your former familiarity will make this quick work." Mr. Pikake stood to draw on the board.

$$x + 6 > 10.$$

"What does this say, boy?"

"x plus six is greater than ten, because the alligator is eating the $x + 6$ meaning it is bigger."

With a subtle snicker, the professor continued. "A large part of the SAN is the study of Numberfolk behavior. Naturally, classifying them is key. We may not always have equalities,

$$41 - (17 - 87) - (-99) - (42 - 73)$$

sometimes the clues may be less obvious. We don't know who is hiding behind x, but," two pops, "we know something about them. Like here, when they are added to six, the result is always larger than ten."

"Okay," Marco began, "but that doesn't really tell us about x, or whoever is hiding behind it. It tells us about $x + 6$, right?"

"Right you are! You can use the same hunting algorithm for duels with inequalities as you can with duels for equalities. Tell, me, how would you isolate the mask?"

This was an easy question. Marco had mastered this dance. "You send in 6's evil twin to duel on both sides." The professor added to the board.

$$x + 6 + (-6) > 10 + (-6).$$

Looking to Marco he bowed his head and said, "Continue."

"Well, on the left you have $6 + (-6)$. So, we've vanquished the 6 and it's simplified to $x + 0$, which of course is just x. And on the right, you have $10 + (-6)$." The ten shot off ten pluses and the six returned fire with their minuses. Fireworks exploded and only the four positives remained. "And on the right, you have four."

$$x + 0 > 4$$
$$x > 4.$$

"Perfection! You see, son, now we know that the Numberfolk hiding must be more than 4. And while not as powerful as hunting and identifying exactly who is disguising themselves, we certainly know much more about them."

In an overcoat, Marco surveyed the scene picking up clues. 'I'm not sure who committed the crime, but I can tell you this!' He removed the magnifying glass from his eye, 'They were at least a five-year-old!' He burst into giggles. *This might be my best daydream yet.*

$$(84 - 24) - (22 - 82) - (16 - 138)$$

"Care to share?" Mr. Pikake prodded hearing his student's laughter.

"I was imagining I was a detective and found $x > 4$. So, it must have been a five-year-old or older who committed the crime." Marco was still laughing.

"But are they?" The professor asked in a dreary tone, pretending to hold his own spy glass.

"If it's greater than four, it has to be at least five, or six, or seven!" Marco snorted back.

"What about four and a half?"

"Oh, I forgot about fractions!" Marco had been enjoying their exploration of the Integers.

"This provides the perfect segue for tomorrow." He stood and walked to the door. "Tomorrow," he looked over his shoulder and flashed Marco a crazy smile, "we shall obliterate!"

15

THE VINCULUM GAMES

"I KNOW NUMBERS ARE BEAUTIFUL. IF THEY AREN'T BEAUTIFUL, NOTHING IS."

-PAUL ERDOS

Numberville followed the patterns of all great civilizations – or perhaps they invented the patterns that all would be doomed to follow. Either way, Natural always had a superiority complex. They saw themselves as better, greater than TNZ and in fact, they were. If they assumed this, then it must also be true that the larger residents possessed a special nobility as well. Somehow, darkness is always more powerful than light. As vast and strong as the sun is, even a tiny and seemingly insignificant ant can cast a shadow, and so it did. The vitriol spread quickly throughout the land, with Numberfolk as small as 5 wielding the little control they had over their neighbors 1, 2, 3, and 4.

As one would expect, social climbing soon became conventional. It was not uncommon for residents to willingly engage in duels to increase their status. If a four and a five battled, the result would be a nine allowing the Numberfolk to enjoy a higher place in society together. While the practice was frowned upon by many, resident after resident willingly left their families behind to combine with another and gain a taste of the high life, well, higher life.

In TNZ, things became hectic. Everyone knew a -1 was larger than a -5 but some insisted on behaving opposite to Natural and building the class system on who was *smallest*. Residents of TNZ were so pessimistic, negative, that they couldn't agree[*].

Some Numberfolk would spend their entire lives climbing. Nicknamed Jumpers, these residents would target oppressed neighbors and convince them to join together. Legend has it that a

---

[*] Very few know, but this is how the concept of absolute value was born. Absolute value considers only the number of particles and cares not whether the particles are negative or positive. Using absolute value $|-5| > |-1|$ since -5 has five negative particles and thus $|-5| = 5$, while -1 has only a single negative particle and thus $|-1| = 1$. The "Absolutes" were the residents of TNZ who insisted that the number of particles were what mattered, not their size. Unsurprisingly, not a single -1 found themselves among the Absolutes.

one made it all the way up to a two million and forty-two through jumping. This cannot be confirmed.

To make matters worse, for some time, faith in the Thinkers had been waning. The elite group seemed to always be reactionary, responding to crises rather than making new discoveries about the power the Numberfolk must have as children of the charmed ones. This allowed a new group to ascend from the shadows, and they called themselves Visionaries.

Amidst the class struggles, Visionaries first arrived on the scene offering residents a new and tempting ability: proliferation. Harnessing Numberfolk magic, Visionaries could offer the power of multiplication. Jumping, a four could duel with a five, but for a price, the same four could partake in an enlargement spell by a factor of five. The result? A powerful twenty that towered over the wimpy nine.

This caused fear and panic in the larger residents. The ability to proliferate, to enlarge, could ultimately lead to the loss of their own standing. Not wanting to join the Visionaries or the Jumpers, the larger residents needed a way to beat them at their own game. They needed something that would distract all of Integer – keep them occupied and take their attention away from the class struggles. The Vinculum Games were born.

The games began as a form of entertainment, a distraction, for the residents of Integer. An exciting diversion that had every Numberfolk cheering and betting on the winning team. Unfortunately, not a single resident had any idea of the colossal impacts a simple game could have. All in all, it was the Vinculum Games that ultimately ushered in the darkest time in Numberville history.

THE STADIUM

The first task was the construction of an arena. The Numberfolk needed to decide where the games would take place. Desiring an inclusive event that would bring together the residents of Natural and TNZ, a team of explorers were nominated to begin scouting for land. They sought to locate a considerably large area to encourage as many spectators as possible. Since Integer was already bursting with Numberfolk and their homes, they would have to travel farther, to a place that had not yet been explored.

Advancing far beyond where any Numberfolk had ever voyaged, the team left the state of Integer and entered the vast and barren country of Rational. As both Natural and TNZ existed within Rational, this was the perfect place to lay their claim and begin construction. The team immediately broke ground and initiated the assembly of the massive game arena.

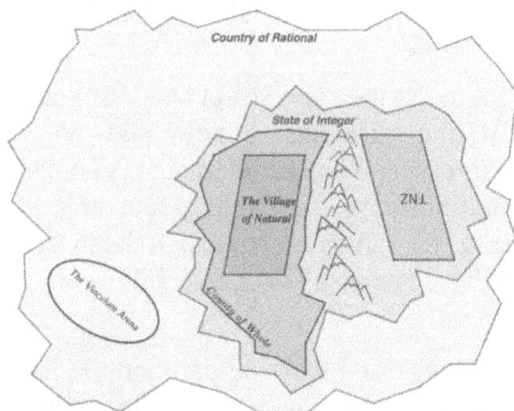

Ovular in shape, the ground floor consisted of the game field. A horizontal line – the Vinculum – divided the playing area into two sides. Each partition contained a gigantic box, as well as four enormous launchers that were equally spaced across the court.

Around the field, on all sides, the team raised stadium seating that seemed to stretch far beyond the sky and ensured as many Numberfolk as possible could attend the games.

THE RULES

Numberfolk viewed players of the Vinculum Games as warriors, gladiators. Being chosen to play was a great honor that resulted in fans and admirers swarming the participants. Aptly, the team stationed on the North Field was called the Number-ators. The name meant to represent the Numberfolk-gladiators who would make their supporters proud. The team stationed on the South Field was dubbed the Denomin-ators. A title that suggested their players were the warriors that would dominate their competition.

A weekly lottery occurred to pick the captain of each team. Once the captains were identified, they would set forth to recruit twenty Numberfolk that would make up their team. At first, the teams overwhelmingly consisted of small Naturals (no one over 100 played in the inaugural games). As play continued, most teams agreed that a mix of residents of Natural and residents of TNZ was the best way to configure a strong team.

During the game, each side would attempt to get their chosen players into the opposing team's box. The game ended when both Number-ators and Denomin-ators had a full box of five players. At this point, the players in the box would duel until only a single Numberfolk, called the Boxer, emerged.

To determine the winner of the game, the Number-ators' Boxer and the Denomin-ators' Boxer would traverse to the center of the field: the Number-ator Boxer above the Vinculum and the Denomin-ator Boxer below. The Great Scale would declare if the result was greater than or less than 1. A result of more than 1 and the Number-ators were the victors, a result of less than one and the Denomin-ators were declared as the winners. If the Great Scale revealed the result was exactly 1, the game was announced as a tie.[*]

After the captains chose their teammates, they would assign each player one of four roles: Blockers, Flyers, Runners, and Slingers. Four Slingers per team would operate the launchers. The launchers were essentially gigantic sling shots that would hurl Flyers at the opposing team's box. A good Slinger could aim so well that the Flyer landed directly in the box like a golfer scoring a hole-in-one.

---

[*] It is important to note that no one had any idea how the Great Scale declared the victor. In R&D (short for Research and Development) when they tested out the game, the Great Scale showed either > 1 or < 1 and they ran with that.

Needless to say, the Flyers were the Numberfolk who consented to be tossed through the air. The Runners' goal, like the Flyers', was to gain entry into the opposing box. Rather than propelling through the air, the Runners stayed low, dipping and dodging across enemy lines.

This leaves us with the final position, the Blockers. Blockers had a simple enough job, guard the box. This meant ramming Runners and intercepting Flyers left and right. However, the Blocker's role was deceivingly complex. These players were required to have the long game in mind. Since the ultimate goal was based on the Great Scale, Blockers needed to determine which players to let into the box and what players to focus on stopping to support their team's end game.

The Vinculum Games commenced and were an instant hit. Trillions of Numberfolk trekked to the stadium each week to watch the games. Some residents became so immersed in the sport that they formed clubs to discuss how to select the best team, strategies to ensure victory for either the Number-ators or the Denomin-ators, and even created fantasy leagues where they would play out pretend scrimmages and declare victors.

However, there were a number of problems that soon emerged. The first: no one had any understanding of how the Great Scale determined if the Boxers resulted in more than or less than one. This made strategizing particularly difficult. The next issue was that no one was sure what happened to the teams. Residents would make flyers and banners celebrating team members, but not a single team member (except captains who could choose not to play and any players who didn't make it into the box) ever returned from the games. The final problem, which was so shocking no one dared even talk about it, didn't occur until the Fifty-third Vinculum Game.

$$\frac{-5000}{-20}$$

THE FIFTY-THIRD GAME

The games were going strong. With little else to do, the pilgrimage to the Vinculum Arena was a national pastime that every Numberfolk had done at least once. On this day, no one, not even the most prestigious Thinkers, could have predicted the outcome of the game.

A 999 was picked to be the captain of the Number-ators and a −68, the captain of the Denomin-ators. Both had selected strong teams. The Number-ators were the first to score. A 21 held the position of Runner. Starting on the far left, she dodged across the Vinculum, and between the opposing team's launchers. At that exact moment, the Denomin-ators' −200 was hurling a 6 which nearly collided with 21. She ducked and slid narrowly missing the Flyer. Faking a move to the left, 47, playing for the Denomin-ators', attempted to block, but 21 was too sly. She spun and whisked herself back to the right before diving for the box.

The 6 made it to the Number-ators box, but not without incident. The close call with 21 sent the 6 tumbling through the air landing on the far right of the field. Luckily, −200 had such strength, 6 landed behind the box which was a rare sight. He was able to easily slip around to the front and claim a goal for the Denomin-ators.

While all this was happening, simultaneously a 3,000, launcher for the Number-ators, propelled a −15 directly into the Denomin-ators box. All eyes were on −15 as they spun through the air. The crowd erupted in cheers as the Number-ators scored another goal. They almost missed the sly −4, also playing for the Number-ators, who had sprinted directly perpendicular to the Vinculum and easily made it into the box. Just as the spectators died down, they jumped up again. The Number-ators had three goals, only two more to go!

The next few minutes saw the Denomin-ators score three successive goals. The −200 sent a −12 through the air while, like his shadow, a 57 ran a parallel path on the ground. Three Blockers ran to intercept the −12 deciding to let the 57 pass. They jumped,

$$\frac{296}{3} + \frac{457}{3}$$

dove, and swatted, ending up in a pile on the ground as −12 soared and safely landed directly in front of the opposing team's box.

That 999 was a cunning captain. She knew misdirection was a strong strategy to get her last players in. Now, mind you, at this time very few Numberfolk actually understood the Vinculum Games. Some had suspicions, but not a single resident could claim they knew for sure. The one thing they had surmised was that since the Number-ators needed the Great Scale to result in 'greater than one' (the scale had never actually revealed the final result – only the winner), their best chance at victory was to have a group in the box that was as close to Zero as possible. The captain had studied previous matches and noticed games that ended with 34 over 2… 476 over 15… and 88 over 12, all had the Number-ators declared as the champions. She took this to mean that making the Denomin-ators' Boxer as small as she could, but positive, was the best play. Seeing this as her chance to make her mark on the history books, 999 had both the players and the smarts to pull off something that had never been done before.

She motioned to the 211 to line themselves up on the right and the −213 on the left. She then directed the −94 and the 132, all Runners, to head towards midfield. When she gave the signal, all four players took off. They leaped across the Vinculum heading towards the box.

As −68 was in his own right a good captain, he commanded all his Blockers to the center to create an impenetrable line. The −94 and the 132 put on an amazing show. Not only did they gain the attention of the crowd but also drew the Denomin-ators' Blockers to them. They zigged and zagged, working together they went around one of the center launchers and seemed to be heading to their side of the Vinculum before doubling back. The Blockers ran to catch the Numberfolk. One went left, the other right, but when the two Number-ator players crisscrossed, they sent the Blockers directly into each other. More Blockers, noticing what was happening, ran towards the center in time to see −94 jump up and flip over the defensive line. In groups of three the Blockers surrounded each of

$$\frac{252}{10} + \frac{1134}{5}$$

the Runners, closing in on them until both were captured and thrown back across the Vinculum.

While all this was happening, it was quite easy for 211 and −213 to slip along the outside of the field quietly making their way to the box. It was 211 who arrived first, staying low and barrel-rolling in to avoid making a scene. The Denomin-ators' Blockers had just thrown their teammates over the Vinculum when −213 approached the box. The only player nearby was −200. With the Blockers busy near the middle of the field, −200 left their post as Slinger and burst towards the box in an attempt to stop them. It was too little too late.

Having five players secured, they proceeded in normal play to duel to determine their Boxer. The Denomin-ators continued to attempt to score to reach their five and were able to slip in a 72 and a 97 concluding the game.

The crowd sat silent waiting for the Boxers to appear. When 220 emerged from the Number-ators box (the result of the $6, −12, 57, 72$, and 97 the Denomin-ators had scored) the stadium burst into a celebration so powerful, even the residents who had stayed home in Integer could hear the screams. Everyone turned to focus on the second Boxer, a synchronous gasp as if the crowd had sucked every bit of air out of the stadium rang through the arena as a Zero stepped into the light. They couldn't believe it. It had never been done before. As the shock subsided, the stands began to shake as in unison every Numberfolk in attendance chanted, "Nine – Hun – Dred – Ninety – Nine. Nine – Hun – Dred – Ninety – Nine. Nine – Hun – Dred – Ninety – Nine…"

Proud of her accomplishment as team captain, 999 stepped onto the field and waved at the crowd. They shrieked in delight. The two Boxers made their way to the center of the court: 220 took his place above the Vinculum and Zero below. Suddenly, a loud tear ripped through the stadium. It was a cross between nails screeching down a chalkboard and the sonic boom of a military jet breaking the sound barrier. Numberfolk were looking around manically trying to determine where the sound had originated. Before anyone could

$$\frac{301}{2} + \frac{615}{6}$$

place it, a gust of wind tore through the stadium knocking spectators into their seats, then another in the opposite direction, and finally the last unnerving gust – this time from below. Numberfolk were torn from their seats, floating through the air, flailing and grabbing at anything they could reach. A booming ZZZIIIPP shook the stadium and threw everyone back down seconds before the explosion.

The event has been described by survivors as a physical tear in time and space. As if an enormous invisible giant had unzipped their winter coat to reveal a colossal black hole that began sucking in everything in sight, a massive vacuum cleaner that jerked half its victims upwards and the other half yanked down before spinning into a tiny ball that disappeared as quickly as it had arrived.

A frenzied exodus came next, as any spectator or player who had managed to avoid the giant's meal ran towards Integer. Numberfolk pushed and shoved, some slamming into each other amidst the panic. The crowd unintentionally rammed a fourteen and a negative eight together, emulating a duel, a six arose from the ashes and instantly continued sprinting away from the arena.

Along came the Thinkers. They were tasked with publishing a report on the events of the Fifty-third Vinculum Games, as well as finding any answers they could about the many lingering questions that haunted some residents[*].

Residents of Integer,

The purpose of this report is to shed light on the events of the Fifty-third Vinculum Games. Our findings may be disturbing to some Numberfolk. We recommend reviewing this news with friends and family before deciding on how best to inform any numberbabes in your care.

---

[*] It is unclear why the Thinkers decided to form this pamphlet as a letter. Some spectate that their wanning popularity led to new approaches to their publications – more friendly.

$$\frac{356}{6} + \frac{423}{3} + \frac{483}{9}$$

We begin with the Vinculum itself. Understanding the function of this device was a key turning point in our investigation. At its core, the Vinculum can be thought of as a divider. It essentially splits the particles of Numberfolk into pieces. Should a 24 be placed above the Vinculum and a 6 below, the result would be a 4. The device split the 24's particles into 6 equal groups, each group having 4 particles each.

Please prepare yourself for our next findings. Residents are all aware of multiplication, be it to produce a child or the banned practice of proliferation to enlarge oneself. The result has always and will always be another resident of Integer. This however is not the case with the Vinculum. Our investigation led us to the outskirts of Rational where we found an entire society of Numberfolk. They were shattered, ripped apart, and dealing with the toll the Vinculum brought. Fellow residents – we are not alone. For the civilization we located is comprised entirely of non-Integers.

Many are likely wondering how this is possible? As we know, all Integers are comprised of particles which can be grouped in various ways. Our test subject, 24, can be split into two groups of 12, or three groups of 8, or four groups of 6, but not all Numberfolk possess this type of flexibility. Consider a 13. It cannot be split – it can be fractured into singletons but split otherwise it cannot. If we used clusters of two, we would arrive at twelve or fourteen, and clusters of three could make twelve or fifteen. Clusters of four can make twelve or sixteen, and clusters of five can make ten or fifteen. You see, there is no way to split a thirteen – or so we thought.

Until now, all we knew of were Integers. While thirteen cannot be split into an Integer, it can be split into something else, something new. We are all comprised of ones and we believed One was whole – it was singular and

$$\frac{206}{4} + \frac{203}{2} + \frac{510}{5}$$

complete. In fact, One can be split. For as large as we are and as vast as we span, there is an entire equivalent universe of parts within each of our particles. And so, while thirteen cannot be divided into wholes, they can be divided into parts. Splitting thirteen in two would result in two groups of 6 whole particles making twelve together, the final particle is then also torn in half, and each part is given to a group, forming a new Numberfolk, one that is not an Integer, $\frac{13}{2}$.

We have named this new civilization the Rationalites and define them as such:

A **Rationalite** is a protected class of Numberfolk created by the Vinculum. Composed of the tearing from the particles of Integers a Rationalite is formed when two Integers are divided by a Vinculum. Rationalites can be partitioned into two classes:

- Integers
- Logos

Let p and q each be Integers, then when combined with a Vinculum p/q is a Rationalite. If the Integer p is built of clusters of q, then the Rationalite is also an Integer. Consider 10, which can be constructed of two clusters of 5, we see that 10/2 is simply the Integer 5. Hence 5 is both an Integer and a Rationalite. What should happen if we instead attempt to divide 2 into ten parts? The Numberfolk 2/10 is created, a Logos. We call this resident two-tenths, and they are both a Rationalite and a Logos, but are not an Integer.

We urge residents to carefully consider the ramifications of these findings. The Thinkers ask all Numberfolk to never forget the atrocities of the Integer Wars and recommend any attacks on these new Numberfolk to be deemed unacceptable. Afterall, they are pieces of ourselves created

$$\frac{642}{7} + \frac{651}{7} + \frac{499}{7}$$

from the Vinculum Games. Furthermore, while not all Rationalites are Integers, all Integers are also Rationalites. We have a natural connection to these new neighbors.

With a refreshed understanding of the Vinculum, we turned our focus to the events of the Fifty-third Games. The next question we studied was what would happen if a zero was placed above the Vinculum and any nonzero Numberfolk below? This is asking if we split zero into clusters, how many would be in each? The result was an unsurprising zero. That is, $0/q$ where q is any nonzero Numberfolk is ultimately 0.

Finally, we were ready to consider a zero below the Vinculum, a zero in the Denomin-ator side that seemed to be the cause of the catastrophic recent events. We chose not to attempt to recreate the incident thus our conclusions are mainly theoretical. Applying the same logic as before, suppose any Integer was above the Vinculum. In the case of the Fifty-third Games, this was a 220, so we shall use this as our example to ensure their sacrifice is always remembered.

The Great Scale attempted to divide 220 into zero parts. Unsure what to do, they began to test values. On the Natural side, it divided 220 into ten equal parts which resulted in 22. It continued towards Zero dividing into 5 equal parts to find 44 and then into 2 equal parts to find 110. As it neared Zero from 10, to 5, to 2 the result was becoming larger and larger. Simultaneously, the Great Scale did the same from TNZ. It divided 220 into −10 parts, then −5, then −2 and the results were −22, then −44, then −110. The values were getting smaller as they approached zero from TNZ! This divergent rip, appearing that on one side the division into Zero parts indicated a very large number and on the other it indicated a very small number, caused the implosion.

$$\frac{3}{2} \cdot \frac{514}{3}$$

Some members of the Thinkers prefer a different perspective. They find it more suiting to consider the Vinculum in terms of multiplication. In the case of 10/2, they ask themselves, 'how many clusters of 2 are needed to make 10?' As this is clearly 5, it is an alternative approach to the same query. To them, the conundrum of the Fifty-third Games was simplified to the task of determining how many clusters of Zero were needed to make 220. The absurdity of the question nearly drove them irrational. No matter how many zero-clusters they amassed, they could never find anything other than Zero himself. It was an impossible task.

Residents of Integer, the Fifty-third Vinculum Games will forever live in our history. This event was traumatic for our community, and it is important that now, more than ever, we support each other and welcome our new neighbors.

Kindly,

Your Thinkers

THE KIDNAPPINGS

The Thinkers' publication tore through Integer and, like dividing by zero, it nearly ripped the state at the seams. Chaos ensued. Integers themselves became divided by their beliefs as factions were established.

The predominate faction was the Supporters. Most Numberfolk were fascinated with the idea of the Logos and the abilities that came with them. The class system was almost a thing of the past as neighbors could barely recognize each other. Tens would wander the streets as 20/2 or 100/10 or even 50/5. There was no limit on the masks they could wear. Every One quickly joined the Supporters. Before this, ones were at the bottom of the Natural totem pole. Now, they would sneak into the Vinculum Arena and create

$$\frac{30}{7} \cdot \frac{301}{5}$$

new Logos, 1/2, 1/3,1/4 and so on. The uncovering of Rationalites meant there were now infinite residents that were both positive *and* less than one which instantly increased their status.

The next faction to emerge was the Purists. They believed Integers were the only true Numberfolk – having been created by the charmed ones and the Mirror of Wonders. They saw the use of the Vinculum to divide up Integers as unnatural, and the Logos were the filthy result of this process. Much later, many Saints found themselves aligned with the Purists. An Exalted Master, Leopold Kronecker once said, "God made the integers; all else is the work of man," highlighting the Purist's prevailing beliefs. Numberfolk in this faction boycotted the Vinculum Games and at times even became violent, terrorizing the Logos and publishing propaganda pamphlets attempting to persuade Rational to war.

The final faction was none other than the Visionaries. With news of the Rationalites spreading, more and more Numberfolk found themselves wanting to better understand who they are and what they are capable of. The Visionaries saw the Vinculum as a device that gave Integers extreme powers and they wanted nothing more than to harness these powers. Seeking to create a potion similar to proliferation, the Visionaries constructed their own device which they called The Great Divider.

A problem soon emerged. The Visionaries needed to run tests, to experiment, with their tool. With no willing volunteers, the Visionaries targeted unsuspecting Integers, whisking away their neighbors in the dead of night to their secret lair.

The Visionaries' cloak-and-dagger approach resulted in little documentation of what truly happened. Numberfolk did, however, have a story they passed down generation after generation. It was a bedtime tale recited to numberbabes to warn them of the dangers of the Visionaries.

THE FABLE OF THE GREAT DIVIDER

Thirty was a lonely Numberfolk. Always grumpy, he had no family or friends and spent most of his days walking the streets of Natural yelling at numberbabes who were not following the rules. Of course, it was no surprise that when he disappeared, he wasn't missed by anyone.

After making his normal rounds about the village, Thirty was only a few paces from his home when a group of Numberfolk grabbed him. He awoke in a dark and unfamiliar room. All around him, the walls were lined with tiny cages filled with kidnapped residents from TNZ.

As Thirty examined his surroundings, he found that he too was caged. Unlike the others, Thirty lay in a large octagon in the center of the room. Bars shot up on all sides. The only exit appeared to be a small door in the ceiling of the cage, much too far for Thirty to reach alone. He screamed and banged on the cage with all his might, but no one could hear him.

Unexpectedly, one of his kidnappers appeared outside the ring. Thirty begged and pleaded with the masked abductor but was ignored. Affixed to the side wall was what looked to be a scoreboard. The kidnapper carefully drew 30 on the left and -5 on the right, before turning to look directly at Thirty revealing an unnerving grin.

Scared and not knowing what would happen next, Thirty was startled by a loud squeak. He looked up to see the door above the cage had slid open. This was his chance! He grabbed onto the bars and furiously tried to climb. The kidnapper laughed as he watched Thirty fall to the ground over and over again.

The shriek of a Numberfolk made Thirty look up. Above the door stood a -5 who was shoved into the ring before the door slammed shut. Thirty immediately understood, he had heard of these underground fight clubs before. He was being forced to battle. Believing it was his only chance to be let go, Thirty dueled.

$$\frac{52}{7} \cdot \frac{385}{11}$$

When 25 emerged from the dust, the kidnapper placed one long mark on the scoreboard. Before he could reason what was going on, another −5 was thrown into the ring. They dueled. A 20 emerged and another mark went on the board. In came the third −5. Another duel and another mark. 15 stood in the ring, tireless, hopeless.

−5. Duel. 10. Mark.

−5. Duel. 5. Mark.

Where Thirty had once stood was now only a Five. He knew this was the end, he had almost nothing left. Like clockwork, the final −5 dropped into the ring. The twins stood looking at each other, neither wanting to move. Both understanding their fate, rather than duel – they hugged. They squeezed each other so tight that the two broke into a cloud of dust, vanquished. A Zero lay on the floor of the ring. The ghost of the Thirty he used to be swirling through his mind. It was six rounds of duels with −5, six marks stood on the board. With a long hook, the kidnappers snatched the Zero and threw him into a smaller cage.

Some say Zero still lays in the cage, constantly begging for freedom. Others claim that he was eventually released to Whole and can still be seen wandering aimlessly and alone. All agree, it was his isolation that caused him to be a victim of the Great Divider. Be kind, surround yourself with family and friends, or you may just be the next victim of the Visionaries*.

---

* The Great Divider proved to the Visionaries what they had suspected. Division was nothing more than glorified subtraction. You see, 30/5 can be viewed as the number of duels required with a −5 for 30 to be vanquished – turned to zero. Since $30 + (−5) + (−5) + (−5) + (−5) + (−5) + (−5) = 0$, the Visionaries found it took 6 duels to vanquish the Numberfolk and thus, $30/5 = 6$.

$$\frac{45}{2} \cdot \frac{58}{5}$$

THE END OF AN ERA

The torture ring known as the Great Divider took many unsuspecting Numberfolk before the operation was eventually disbanded. Some believe the military term "Zero-dark-Thirty" is a reflection of this tale. It describes the darkest hour, 12:30 a.m., to reflect the darkest time in Numberfolk history. The Visionaries found what they needed, and Suppression was born. Although not as popular as Proliferation (it was rare for a Numberfolk to want to become smaller in size), Suppression was a way to lessen the number of particles. The real power of Suppression however, was that it now gave Numberfolk the ability to turn themselves into anyone – including turning an Integer into a Logos.

Proliferation only allowed Numberfolk to change into what we call multiples of themselves. It was a cloning spell. A seven could only become $14, 21, 28, 35, \ldots$ The results of $7 \cdot 2, 7 \cdot 3, 7 \cdot 4, 7 \cdot 5, \ldots$ and so on. A seven could not become a ten under Proliferation.

Suppression was the result of multiplication by a Logos. Therefore, a seven under a $1/7$ Suppression charm could become a one! More interestingly, now a seven could become a ten using a combination of Suppression and Proliferation as $(7/7) \cdot 10 = 10$.

Much of what we know today came from the Visionaries. The Thinkers also continued their studies and eventually a full understanding of Rationalites existed throughout Numberville. We summarize some of the key findings below.

1. **Form**. Rationalites have the ability to disguise themselves in infinite forms. This power is known as *reducibility*. For a $\frac{1}{2}$ can appear as $\frac{2}{4}, \frac{4}{8}, \frac{100}{200}$, and so on and so on.

2. **Duels**. Logos can only duel when their Denomin-ators are the same. These duels became a sort of circus side-show for

$$\frac{393}{2} \cdot \frac{4}{3}$$

Integers who were enthralled with the new and intriguing process. If a $\frac{1}{2}$ attempted to duel with a $\frac{1}{4}$ they would simply bounce off each other, unable to combine. Duels between Logos started with each dueling Numberfolk changing their form to have the same Denomin-ator, before sending their Number-ators to duel alone. In the case of $\frac{1}{2}$ and $\frac{1}{4}$, the $\frac{1}{2}$ would alter itself to become $\frac{2}{4}$ before battling. As they approached each other to duel, their Vinculum would expand as both it and the Denomin-ator became one, $\frac{1+2}{4}$. Atop the long bar the Number-ators were now free to duel. The result of the Number-ator duel was of course, $2 + 1$, and thus $\frac{3}{4}$ would emerge from the dust.

3. **Multiplication**. Logos went on to have families and numberbabes like all other residents. Multiplication followed the same structure as with Integers, with the exception of both their Number-ators and Denomin-ators

$$\frac{789}{5} \cdot \frac{5}{3}$$

would multiply. For instance, the product of a $\frac{2}{3}$ and a $\frac{6}{7}$ was

a $\frac{12}{21}$ who would reduce to a $\frac{4}{7}$.

To many, the Logos became known as *Fractions*. This name was chosen to represent the factions that divided Numberville upon their arrival.

As time went on, Rational became swarmed with residents. We might even think that there were more possible Logos than there were Naturals. Afterall, from 1 to 10 there are exactly ten possible Natural residents $(1, 2, 3, 4, 5, 6, 7, 8, 9,$ and $10)$ yet from 1 to 10 there are an infinite number of possible Logos[*]. Surprisingly, both Logos and Natural had the same population.

The country of Rational became an urban metropolis. While Integers didn't know *every* resident of Natural and TNZ, they at least knew *of* everyone. The bustling of new faces worried many, and so the Partnership Program was initiated. The Partnership Program paired *almost* every Integer with a Logos as an attempt to smooth tensions and build bridges between Rationalites within and outside of Integer.

They called these partnerships *reciprocals*. The term was meant to show how each companion would reciprocate, or return, the kindness extended to them. Partners were paired with Numberfolk who had innate similarities. To do this, a masquerade ball was hosted in the outskirts of Rational. Every Integer arrived in a basic, but effective, disguise and lined up anxious to meet their new ally. The theme of the ball was 'over one' as the Numberfolk had costumes that placed themselves, well, over one. The 2 who was

---

[*] This might be challenging to see, here are ten Logos just between $\frac{1}{3}$ and $\frac{1}{2}$!
$$\frac{2}{5}, \frac{7}{20}, \frac{10}{21}, \frac{21}{62}, \frac{99}{267}, \frac{34}{97}, \frac{77}{166}, \frac{37}{99}, \frac{177}{415}, \frac{117}{337}.$$

first in line dressed up as $\frac{2}{1}$, and the -5 who was next came as $\frac{-5}{1}$ and so on and so forth.

Each Integer was placed in a device that would reveal their partner by flipping them on their head like a child practicing gymnastics to indicate their match. Two was giddy when they were flipped to uncover $\frac{1}{2}$ as their partner. And negative five giggled manically when $\frac{-1}{5}$ was unveiled.

The ball was a great success but wasn't without hiccups. When one approached the machine as $\frac{1}{1}$ and flipped to reveal $\frac{1}{1}$, another Integer, they erupted in tears. Neither 1 nor -1 had a reciprocal partner that was a Logos. Panic ensued when the organizers almost let a zero enter the machine. Dressed to the nines as $\frac{0}{1}$, Numberfolk who had witnessed the Fifty-third Games jumped and pushed to stop things just in time before the bursting hole of death known as $\frac{1}{0}$ was introduced.

When the night concluded, every Integer with the exception of $-1, 0,$ and 1 had their pair. Unfortunately, this left many Logos without partners. For instance, the reciprocal of 3/4 wasn't an Integer at all, but another Logos, 4/3. These Logos alongside $-1, 0,$ and 1 banned together to form their own support group and the program was widely considered a victory. Many reciprocal partnerships remained friends long after the program ended, but some fell in love and went on to produce numberbabes. Interestingly, the product of reciprocals was always a 1.

As more and more Rationalites became friends and family, Numberfolk discovered that Logos could be created without the use of the Vinculum or its immoral counterpart, The Great Divider. For instance, suppose the Vinculum was to create a resident with a 3 on the Number-ators' side and a 4 on the Denomin-ators' side. The result would of course be the Logos 3/4. However, Numberfolk found that should a 3 simply multiply with a 1/4, that too would produce a 3/4. They found that dividing by a Numberfolk was the exact same as multiplying with the Numberfolk's reciprocal.[*]

Knowing they could always create new Rationalites through loving families, they saw no need for the sport. The Vinculum Games were abandoned and the stadium boarded up. While Numberfolk loved the games, they always reminded residents of the dark times fueled by the Visionaries and the Great Divider. Everyone agreed it was best to leave them in the past and usher in a new era of Numberville.

While many Numberfolk moved on, Visionaries still lurked in the shadows, becoming skilled at isolation. They knew how to vanquish any resident, simply force a duel between it and its evil twin. But, the Fable of The Great Divider cautioned everyone to avoid being alone – it made Numberfolk smarter and savvier. The new developments and the Partnership Program led the Visionaries to create a new device: the obliterator. The obliterator could turn any Numberfolk into a one by spitting out its reciprocal and binding the two together. As Numberville moved on working hard to put the horrible times behind them, the Visionaries – with their ability to vanquish and obliterate – were able to continue their work and experimentation hidden in plain sight.

Little is known about the true nature of the Spogs and the initial creation of the Integers. But it can be said that Kronecker was at least half-wrong. It wasn't humans who created the Logos, it was the Numberfolk themselves.

KNOWING THE ENEMY UNIT 1
VINCULUM VIRTUOSO

---

[*] In this case, dividing by 4 is the same as multiplying by $\frac{1}{4}$, the reciprocal of 4. And dividing by $\frac{1}{4}$ was then the same as multiplying by 4.

The Vinculum Games were truly a spectacular thing to witness. Numbers flying and running everywhere, the Boxer, and all the tension was simply amazing.

I figure this is a good time to clarify a few things about the games.

First, there were actually two types of game play: normal and tournament style. Normal game play followed the rules presented here. In tournament style, things were more challenging as each team didn't need a full box. The first team to score five in their opponent's box would end the game and determine the winner.

The game is also a bit lopsided. As the Great Scale rules on greater than or less than 1, this means all the negative values and all the values between 0 and 1 end up as a win for the Denomin-ators, and the Number-ators only have the positive values greater than 1.

There is a lot of strategy involved in terms of the plays. You want to keep track of the box to know who to send in. For instance, say you are captain of the Number-ators. The Denomin-ators have sent players into your box which results in a negative value. To win, you want to send players to the Denomin-ators box that result in a negative, but a negative greater than the total in your box. Conversely, if the Denomin-ators have sent players into your box that results in a positive, your goal is to keep the Denomin-ator box positive but smaller than your box.

If you have a chance to play on a fantasy league, we highly recommend it. These mock-battles can become very intense and competitive but are always a ton of fun.

2^4

$10^2(3.14) - 298$

$$\frac{192}{12}$$

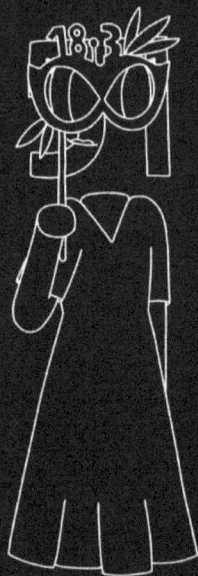

$\sqrt{256}$

$0.4(44 - 4)$

$2^2 + 2^2 + 2^2 + 2^2$

$10\left(2 - \dfrac{2}{5}\right)$

16

OBLITERATE

"MATHEMATICS IS THE LANGUAGE IN WHICH THE GODS SPEAK TO THE PEOPLE."

-PLATO

aturday night was again full of vivid dreams. The idea of obliterating sent Marco's imagination on a wild ride. Hardly able to wait, he arrived at the library thirty minutes early. His face lit up when Mr. Pikake finally strolled into their study room.

"Obliteration!" the professor exclaimed, "The object we are obliged to observe!" He walked to the front of the room before pivoting on one foot to face Marco. He flashed a devious grin, "Ready?"

Am I ready? Marco couldn't wait to find out what obliteration was and how to do it. He nodded enthusiastically and didn't take his eyes off the professor as Mr. Pikake recounted the Vinculum Games, the factions, the Logos, The Great Divider, and the Partnership Program.

"Wow," was all Marco could muster at first. The imagery of Numberfolk hurling through the air and the Visionaries cloaked in secrecy offering up potions to enlarge or suppress their peers was enough to keep his imagination fueled for days. He wanted to make sure he understood everything the fantastical tale had to offer. "So, all the Integers and the Logos are Rationalites?" he asked first.

"Correct!" Mr. Pikake shot his hand in the air. "A Rationalite is simply two Integers and a Vinculum, like $\frac{4}{9}$ or $\frac{17}{3}$ or even $\frac{25}{5}$."

"And $\frac{25}{5}$ is just 25 divided by 5. So, that is 5, an Integer. The others, like $\frac{17}{3}$ are the Logos?"

"Perfection!" the professor purred.

"Okay. And then I know about adding fractions, or, *um*, dueling Logos. You have to make the denominator the same and then you just add the numerator."

"Well done, boy. But," this was the loudest pop Marco had ever heard from the tutor, "you have missed the critical clue! It is important to understand that all the Rationalites, Integers, and

Logos have infinite disguises they wear. Think of the simple resident 2. Now it knows how to disguise itself as $\frac{4}{2}$ or $\frac{6}{3}$ or $\frac{10}{5}$ or $\frac{24}{12}$ or $\frac{100}{50}$ or..."

"Okay, okay, I get it." Marco chuckled stopping Mr. Pikake from continuing on forever. "Multiplication means the numerators multiply and the denominators multiply and the same thing happens with the numberbabes? Like a negative times a negative is a positive."

"Precisely. I suppose the tale doesn't tell us what happens when a negative is in the Number-ators' box, and a negative is in the Denomin-ators' box." He scratched his head trying to remember. "Anyhow, that too is positive for the same reasons. There are many schools of thought. I prefer never to divide as division is simply multiplication by the reciprocal, the buddy assigned in the Partnership Program. And the Great Divider is simply too distasteful for me. If -32 was in the Number-ator and -8 in the Denomin-ator, I would simply see this as -32 times negative one-eighth, $-32(-1/8)$. Thus," he paused and looked to Marco to ensure he was following along, "thirty-two clusters built in the upside-down then the negative one-eighth fuses the clusters in the right-side up. However!" He held the last word like singing an opera, "Because it is a Logos, instead of building eight clusters, it splits the thirty-two clusters into eight groups and the numberbabe is simply one of the groups, one of eight. Whatever you pick, it's a four. Yet, there is much debate about the temperance of the child – perhaps one of the four clusters is more well-behaved than another and so forth."

"And dividing by a fraction, a Logos, how does that work?" Marco chimed in, sparkling with interest.

"Repleted with requests today, *ay*?" Mr. Pikake teased and Marco sheepishly looked down. "Remember, division, whatever lay beneath the Vinculum, is easily equivalent to multiplication by the reciprocal. Incidentally, $\frac{1}{4}$ divided by $\frac{2}{3}$ is simply $\frac{1}{4}$ times $\frac{3}{2}$, which is,"

"Which is $\frac{3}{8}$!" Marco burst in.

"And a master of the Logos you have become." The professor circled his hand before giving Marco a deep bow and sliding into the chair.

"Thank you," Marco replied humbly. "Just one more question. About the reciprocal, the Partnership Program."

"How to obliterate? That is our question! I anticipate it will answer your query." Mr. Pikake stood up, slowly stretching his long limbs before walking to the board. "Numberfolk learned from the Fable of the Great Divider. It is rare to find them wandering alone, they travel in packs, making it challenging to isolate them. On top of this, they use their proliferation and suppression charms to distort their faces. But," a pop, "with the obliteration power, we can tear this all away – strip them down to their basic components!" He slashed at the board.

$$4x$$

"Do you recall what this means?"

Marco did, "It means 4 times x or that there are four x's hiding together."

"Exactly! But we need only one to be able to complete our hunt. Let us review how to vanquish. Why did we vanquish? What was our goal?"

Silence filled the room as Marco contemplated the question. "I guess. I guess the idea was that the numbermask is not alone, they are with others. Like $x + 7$, the mask x is hanging around 7, dueling, and we need to get it alone. We vanquish the 7 to isolate the x."

"Very good. An important part of vanquishing was sending in the evil twin because a duel with an evil twin resulted in a zero.

And a zero has the ability to reflect in a duel. But Zero does not have this ability when masquerading is occurring.[*] If we have zero x's we have lost the mask entirely! We have no hope of ever uncovering it."

Marco was following along. He thought about vanquishing the numbermask, he pointed a magical wand at the equation and POOF! In a cloud of smoke, the numbermask had entirely disappeared, gone forever making it impossible for Marco to finish his hunt. "We don't want zero masks, we want one, right?"

"Right! This is precisely what obliteration does. It changes the number of masks to only one, allowing us to zone in, to attack!" He wielded an invisible sword going in for the final blow.

Marco stared hard at the board. How was it possible to simply change a 4 into a 1? He thought about asking, but Mr. Pikake was already moving on.

"I told you of the Partnership Program. Every Integer has a reciprocal, 2 has $\frac{1}{2}$, -5 has $-\frac{1}{5}$, and so forth. But in fact, all Rationalites have reciprocals. It is as if you turn the Numberfolk on their head, turn them upside-down. The reciprocal of $\frac{2}{3}$ is $\frac{3}{2}$ and of $-\frac{4}{5}$ is $-\frac{5}{4}$. What is marvelous is the product of any Rationalite and their reciprocal is precisely 1. Obliteration is the process that takes their *coefficient* and forces it to become 1."

The word *coefficient* sounded familiar, but Marco wasn't entirely sure what it meant. Sensing it was an important part of obliteration, he asked, "Remind me what a coefficient is?"

---

[*] In case you forgot, *masquerading*, is the process of a Numberfolk warping their mask. It is either through enlargement or suppression a numbermask may appear for instance as $3x$ or $\frac{1}{3}x$ or $\frac{2x}{6}$. All are masquerades, disguising the mask itself.

$$4x = 1092$$

"Excellent inquiry! The prefix 'co' means with, as in the words cooperative or coordinate. To do together, if you will. The coefficient is how many numbermasks are working together – it is the factor of the masquerade, the distortion spell. If the term is $4x$ the coefficient is 4, because there are four of them. If it is $\frac{1}{2}k$, the coefficient is $\frac{1}{2}$, the numbermask has split themselves in two as a disguise. Understand?"

The details were starting to return, Marco remembered learning about the word last year when he had been introduced to terms.

"Now, to obliterate, you simply cast the reciprocal. It will force the numbermask out of hiding and allow us to uncover it. But," pop, "anything we introduce into an equation requires us to equally apply the action, scales and all, both vanquishing and obliteration must be done to both sides of an equality." He added an equation to the board.

$$4x = 16.$$

"This one is easy enough to begin with. Do you know who is hiding behind x?"

Imagining M&M's, Marco arranged sixteen in his head, he broke them into four even groups and fashioned them into X's. Each X was made of two M&M's on each diagonal, four total. Four groups of four was sixteen, "Four!" he blurted out.

"Amazing! Numberfolk are rarely so obvious, but this provides a perfect introduction to obliteration. Once again, to obliterate, we cast the reciprocal. Casting an obliteration spell impacts all."

"Like to duel?" Marco wasn't sure what the man meant by 'cast'.

"*Ah*, not to duel. The 4 and the x are not dueling, they are fused together through Proliferation – an enlargement spell. Obliteration rips them apart, it divides them. And division is simply multiplication with the reciprocal."

$$\frac{8}{137}x = 16$$

Marco wrinkled the left side of his face. He didn't really like the idea of multiplication or division. Multiplication was how numbers had babies and division was basically a torture device. He felt thankful for the Visionaries and their spells as he could simply enlarge or suppress the numbermask with a flick of his wrist and push away these unwanted Numberfolk thoughts.

Checking his understanding he asked, "So we multiply both sides by the reciprocal of the coefficient?" His words surprised him, they were more mathy then he had intended. Marco was starting to sound like a master. Mr. Pikake shot the boy a wink.

"Precisely! And in $4x = 16$ the coefficient of the numbermask is whom?"

"Four."

"And it's reciprocal?"

"One over four. One-fourth, $\frac{1}{4}$."

Mr. Pikake added to the board:

$$\left(\frac{1}{4}\right)4x = \left(\frac{1}{4}\right)16.$$

"The obliteration is obvious. We know what it will do. It turns the 4 x's into a single x.° We need only to compute the right. It is asking, what is one fourth of sixteen or perhaps a more friendly proposition, what is sixteen divided by four?"

The vacuum returned, Marco set it to four and sucked up the sixteen. Doing its job, it ripped the sixteen apart into four groups

---

° Since $\left(\frac{1}{4}\right)4x = \left(\frac{4}{4}\right)x$ and $\frac{4}{4} = 1$, obliteration casts a charm on the coefficient reducing or enlarging the mask to one. Remember Zero and One are special Numberfolk. One has the power to both disappear and, like Zero reflects in a duel, in proliferation One also reflects their partner. $1 \cdot a = a$.

$$\frac{4}{3}x = \frac{1100}{3}$$

each containing four. "Well sixteen divided by four is four." Marco hopped out of his seat and added to the board:

$$x = 4.$$

"Stupendous! Attempt this one."

$$-104 = 8x.$$

Marco readied himself, "Okay. We need to obliterate the eight." He imagined the Vinculum. The number eight was sitting atop the bar when suddenly like a gymnast, he flipped, now hanging below. Then lightning struck, electrifying the bar. Obliterated. Marco chuckled.

"We send in the reciprocal to both sides."

$$\frac{1}{8}(-104) = \frac{1}{8}(8x).$$

"We obliterated the coefficient, so the right is just x, all alone." He snickered. "But I'd have to do some figuring on the left." He scribbled on his paper to divide 104 by 8. When he was satisfied, he completed his hunt on the whiteboard.

$$-13 = x.$$

"Beautiful!" Mr. Pikake smiled kindly.

Finding himself enjoying obliteration more than dueling, Marco brandished his imaginary sword. "Give me another."

With an adoring smile, Mr. Pikake obeyed and scribbled:

$$14b = 70.$$

Marco slashed at each side of the equation. His knowledge of fractions allowed him to skip a step. He knew $\frac{1}{14}(14) = \frac{14}{14}$. So this time, using the marker as a sword, he struck the board quickly.

$$\frac{2}{5}x = \frac{1656}{15}$$

$$\frac{14b}{14} = \frac{70}{14}.$$

Since he obliterated the coefficient on the left, all Marco needed to worry about was the right. He counted by fourteens in his head: 14, 28, 42, 56, 70. There were five fourteens in seventy. With his final slashes, he concluded:

$$b = 5.$$

"Found you!" Marco pointed the marker into the air in celebration.

"Okay, okay!" Mr. Pikake laughed, "I see you have taken to obliteration quite quickly. Let me present a more challenging conundrum."

$$6 = \frac{k}{-2}.$$

Ahh, he is trying to trick me, Marco thought. He saw the Vinculum again, this time the -2 was hanging below, sticking out it's tongue and swinging back and forth. It did a full spin around the bar before landing on top, showing off their perfect trick. Marco wasn't sure if what he was seeing was right, but he thought he'd give it a try.

"The reciprocal of $\frac{1}{-2}$ is -2." He looked to Mr. Pikake for validation before continuing. The slight smile and nod told Marco he was on the right track. "I obliterate with a -2. On the left -2 times 6 is -12 and on the right, well, that is just k, all alone."

$$(-2)6 = (-2)\frac{k}{-2}$$
$$-12 = k.$$

"And they said you were not a math kid!" Mr. Pikake let out a jolly chuckle. "Lad, tell me, how did you know it was $\frac{1}{-2}$? That is a tricky talent of Numberfolk that confuses even the best of us."

$$\frac{2}{3}x = \frac{554}{3}$$

Marco felt proud, "Well $\frac{1}{-2}k$ is $\frac{k}{-2}$! So even though I couldn't see it, I knew the coefficient was $\frac{1}{-2}$. °"

"Amazing. You are becoming quite the hunter, son. These are the tricks that confuse and confound and cause countless to cower." He tapped his finger on the table before commanding, "Add your new skills to your journal."

Marco dug the book from his backpack. He was excited to add to its clean pages. He thought for a minute about what to write to get the words just the way he wanted.

> When multiple numbermasks are grouped (through proliferation or suppression not through duels) like in the case of 4x or ½ k, I can use obliteration to get a single mask alone.
>
> To do this, I first must identify the coefficient. The coefficient is the number attached to the mask, telling me how many masks there are. In 4x there are four masks, in ½ k a mask has been divided in two.
>
> Sometimes the coefficient hides as well. In the case of k/2, I must decode that this is really saying ½ k.
>
> After the coefficient has been identified, I can obliterate. I do this by sending in the reciprocal. The reciprocal is a Numberfolk flipped upside down. The reciprocal of ½ is 2 and the reciprocal of 5/9 is 9/5.

° How did Marco know this? The invisibility power of One! Any Numberfolk that stands alone, is actually standing on a One. That is, 4 is really $\frac{4}{1}$ and so k is really $\frac{k}{1}$. All that is left is the matter of multiplication. $\frac{1}{-2}k = \left(\frac{1}{-2}\right)\left(\frac{k}{1}\right) = \frac{1 \cdot k}{-2 \cdot 1} = \frac{k}{-2}$.

$$\frac{5}{2}x = 695$$

Obliteration, like vanquishing, must be done to both sides to maintain the scales.

"I think a change of scenery is in order?" A smile spread like butter over Mr. Pikake's chin.

The two gathered their things and pushed the library doors open to be smacked in the face by winter's bitter, icy breath.

"Snowball fight?" The tutor's eyes were fiery and wicked.

Marco dove behind a bush in the small courtyard for cover. Tossing his briefcase to the side, Mr. Pikake ducked behind a bench and began crunching together piles of untouched snow into firm spheres.

"You see $3y$," he shouted in Marco's direction before throwing his hand up and to the side, like a catapult launching the frozen orb in the boy's direction.

"The coefficient is 3, I need to slash it into thirds." Marco ducked, avoiding the projectile before flinging his own. "Divide by three to get y alone!"

"Nice!" Mr. Pikake popped his head up just in time for Marco's ball to hit him straight in the face. He laughed and brushed himself off. "Round 1: name the reciprocal. Round 2: hunt."

"Fair game!" Marco shouted.

"Reciprocal of 6?" He dodged to the left to avoid the incoming missile then returned fire.

"One sixth, $\frac{1}{6}$" Marco was hit, he fell back into the snow, rolled for cover, and started building up an arsenal.

"How about of $\frac{4}{7}$?"

"Seven fourths! $\frac{7}{4}$." The two were both frantically building as many balls as they could.

$$\frac{x}{9} = 31$$

"And of $\frac{3}{2}$?"

"Two thirds! $\frac{2}{3}$."

"Okay, okay, I can see you have the idea. Ready for Round 2?" The tutor peeked over the bench, ready to let loose.

"Go!" Marco shrieked. As if the flood gates opened, they began frantically pitching snowballs at each other. Too quickly to even try to duck or avoid, they were both instantly covered in snow. As they threw, they yelled back and forth.

"Two-thirds x equals four!" Mr. Pikake shouted.

$$\frac{2}{3}x = 4.$$

"Multiply both sides by three over two. The two-thirds is obliterated and four times three is twelve and twelve divided by two is six! So x equals 6."

$$\left(\frac{3}{2}\right)\frac{2}{3}x = \left(\frac{3}{2}\right)4$$
$$x = 6.$$

"Six x equals 5!"

$$6x = 5.$$

"Easy! Slash both sides by 6, $x = \frac{5}{6}$."

"Four y over nine equals seventy-two!"

$$\frac{4y}{9} = 72.$$

"*Uh,*" Marco stopped for a moment, long enough for Mr. Pikake to get him right on the chin. "No fair!" he laughed. "Okay. Obliterate with nine over four. That's too hard!"

"I'm out of ammunition anyway," Mr. Pikake chuckled and stood, brushing himself off.

$$\frac{1}{8}x = 35$$

"I think I'm actually frozen." Marco struggled to move, his knees locked from a combination of the cold and kneeling for so long. The professor walked over and helped him up. "Seventy-two times nine over four is the same as seventy-two divided by four, which is eighteen and then eighteen times nine which is one hundred and sixty-two." He wrote the problem out in the snow.

$$\frac{4y}{9} = 72.$$

"What do we do first?" he looked to Marco.

"The numbermask y is hiding using $\frac{4}{9}$ of their mask. We find the reciprocal, which is $\frac{9}{4}$ and fuse that – multiply it – to both sides."

$$\left(\frac{9}{4}\right)\frac{4y}{9} = \left(\frac{9}{4}\right)72.$$

"Keep going," Mr. Pikake urged.

"Okay. On the left, we obliterated the $\frac{4}{9}$ because nine fourths times four ninths equals one. That means whatever $\frac{9}{4}$ times 72 is, is hiding behind y."

$$y = \frac{9}{4}(72).$$

"Very good. But who is this?" The professor was clearly not going to accept Marco's attempt to skip the computation.

"Alright, alright. Seventy-two fourths is 18 and eighteen times nine is 162." Marco had cheated, he didn't actually do the calculation, he just remembered what the professor had told him. However, he was tempted to give Maggie's distributive strategy a try by changing the question from 18(9) to (10 + 8)9. Then, he would have 90 + 72 which is 162! Catching on, the tutor flashed Marco a look before adding, "I see what you did there."

$$\frac{5}{3}x = \frac{1405}{3}$$

"Actually, I changed eighteen times nine to ten plus eight and that whole thing times nine to find ninety plus seventy-two." It was a lie, but a small one. He had actually done that, *eventually*.

"I am very impressed! Not all masters are expert computationalists. That is a great skill indeed!"

Now, Marco started to feel a little badly. Afterall, he hadn't really done the calculations at first and it was Maggie's trick. He tried to push the feeling away as he leaned to see what the professor was doing on all fours.

Mr. Pikake was scribbling battles in the snow. "Alright, advance your ability!" He floated his arm in the air, his palm towards the sky, as he presented Marco the writings. And Marco hunted. He hunted each and every numbermask with grace.

$$13x = 78$$
$$\frac{1}{13}(13x) = \frac{78}{13}$$
$$x = \frac{78}{13}$$

$$-\frac{14}{2}y = 28$$
$$\left(\frac{-2}{14}\right)\left(-\frac{14}{2}y\right) = \left(-\frac{2}{14}\right)28$$
$$y = -4$$

$$\frac{3k}{8} = 33$$
$$\frac{8}{3}\left(\frac{3k}{8}\right) = \frac{8}{3}(33)$$
$$k = 88$$

$$\frac{8}{9}l = 82$$
$$\frac{9}{8}\left(\frac{8}{9}l\right) = \frac{9}{8}(82)$$
$$l = \frac{9}{4}(41)$$
$$l = \frac{369}{4}$$

$$\frac{x}{6} = 47$$

"Superb, son! Simply spectacular! Although, one Numberfolk still hides behind a mask."

Marco studied his work. It was already starting to fade as their body heat was melting the snow. He knew it wasn't the final hunt, 4 was only made of twos which 369 could not be built with. It must be 78/13. He counted in his head: 13, 26, 39, 52, 65, 78. In shock, Marco quickly used his finger to add $x = 6$. "That was a good mask for s-s-s-i-x," he blurted.

"I don't think I have told you how proud I am, son. You are simply spectacular!"

Marco was shivering, "T-t-thank you."

"It is imperative you get home and warmed. I must impart you with important information before you go!"

"O-o-okay."

"Recall, our inequalities, when we have clues about the Numbermask, but cannot outright hunt them."

"Y-yes."

"You may obliterate these as well! With one crucial condition!" He bent down and, in the snow, wrote:

$$4x > 8.$$

"Here, we obliterate the four by casting the suppression of $\frac{1}{4}$ to each side." He added:

$$\left(\frac{1}{4}\right)4x > \left(\frac{1}{4}\right)8$$
$$x > 2.$$

"G-g-good. Easy to remember, same thing." Marco's teeth where chattering beyond his control.

"Yes. Except, when there is a negative. Look!"

$$3 > 2.$$

$$3x = 849$$

"I-I-I don't get it, three-e-e is gr-r-reater than two and there are no-oh-oh negatives."

"Ah, but if we multiply both sides by $-1...$"

$$-3 > -2^*$$

"we no longer have a true statement."

"So-o-o what do-o-o you do-o?"

"You must flip the sign. If you obliterate by a negative in an inequality, since you are basically changing everything from Natural to TNZ or vice versa, you need the opposite sign." Mr. Pikake erased the $-3 > -2$ with his foot before replacing it with:

$$-3 < -2.$$

"O-opposite sign for o-obliteration by negatives. G-g-got it."

"Okay, okay. Get home, son." He reached into his briefcase and pulled out a slip of paper. Marco kept both arms wrapped around his body turning to the side to allow his hand to meet his homework. "Hunting practice before next weekend. Then, we shall step it up to complex scenarios!" He raised one arm into the air as he pivoted and ran off in the opposite direction. Although much better at hiding it than Marco was, he could tell the professor was also freezing by the way his long legs bounced in leaps.

Marco laughed at the sight before doing his own sprint back home.

HUNTING 101 UNIT 2

---

* Notice this is not true! When you multiply or divide both sides of an inequality, the inequality sign must flip – change directions.

$$\frac{1}{4}x = 71$$

It can be easy to forget that we can treat inequalities almost exactly the same as equations. You can vanquish and obliterate, with the only exception being that pesky negative.

When we multiply or divide by -1, we are essentially flipping the number line on its head. For example, if $x > 3$, multiplying by both sides and *not* flipping the sign would give us $-x > -3$. Is this right?

If $x > 3$ that means x could be 4 or 5 or 6 and so on and so forth. It could also be all the Numberfolk in between each Natural. What about $-x > -3$? If $x = 4$ then $4 > 3$, but notice that $-4 \not> -3$. When we flip the sign to obtain $-x < -3$, all is good in the world.

You can relate this idea to Geometry. A reflection in Geometry is what happens when we flip a shape over a line. It is the same thing you see when you look in a mirror. You might raise your right hand - but mirror-you is raising their left. Creepy.

Multiplying by -1 is like being sling shot over the mountain dividing Natural and TNZ. If you started as positive, a good dose of -1 will make you negative. If you started as negative, a cup o' -1 is just what you need to feel positive again.

Since everything in TNZ is the opposite of Natural and vice-versa, this means inequalities are opposites too. In TNZ -6 is *less than* -4 so $-6 < -4$. For their twins in Natural it is the opposite, $6 > 4$.

All in all, if you throw your inequality over the mountain, make sure to account for the fact that you have stepped through the mirror into opposite-land. I wonder, if I am right-handed in the normal world, does that mean I am left-handed in the mirror-world?

$$\frac{x}{19} = 15$$

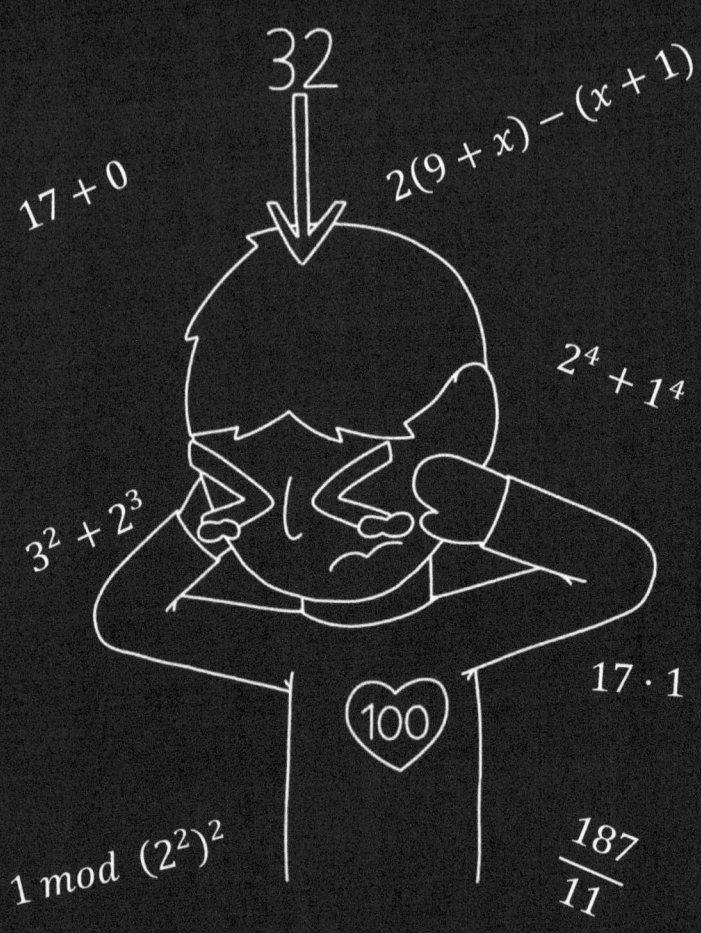

17

MARCO INTERRUPTED

*"ALTHOUGH HE MAY NOT ALWAYS
RECOGNIZE HIS BONDAGE,
MODERN MAN LIVES UNDER A
TYRANNY OF NUMBERS."*

-NICHOLAS EBERSTADT

It was rare for Marco to look forward to a Monday. Mondays meant waking up early. Mondays meant school. Mondays meant homework. Worst of all, Monday was the start of five consecutive days of the same. But, this Monday was different.

It was the last week of January. Winter could still be felt in the air, despite the days stretching slightly longer, the sun shining a little more, and the feeling that with a pinch of luck, spring would arrive soon. In all fairness, it wasn't Monday that excited Marco. It was Thursday. Monday just meant it was a little bit closer. Thursday was the first game of the season, and Marco couldn't wait to get back on the field with his team.

When Marco arrived at school, he was surprised to find a small yellow slip of paper folded and taped to his locker. It read, 'Lunchtime Meeting – Mrs. Sanders'. Marco was upset. Why on Earth did his math teacher need to see him? Not only was he doing much better in class, it was ridiculous to take away his lunch break on a Monday!

It was amazing how one little note could ruin his entire morning. Everything was a slow daze as Marco turned idea after idea over in his head trying to figure out what this meeting could possibly be about. Marco was so far from reality, Liam's lunchtime explanation on aliens and how they are already among us didn't even sink in.

"Marc? Earth to Marc!" Liam waved his hand back and forth in front of his friend's face. "You thinking about which of the teachers are aliens? It's crazy right?"

"No. No. I have a lunchtime meeting with Mrs. Sanders, and I have no idea what she wants to talk about. It's been freaking me out all day."

Oliver motioned to the clock, "You better get going or you'll miss it. Wouldn't want the teacher's pet to be late."

Marco groaned before jumping up and sprinting down the hallway. He flung open his math teacher's door and stopped in his tracks. By the window, he saw a long slender body. Dressed in gray slacks and matching suit jacket, the man's legs pointed towards the door as he twisted his torso to look out the window at the front parking lot. Marco couldn't see his face, but he'd recognize the spider-like features and unkept hair anywhere. "Mr. Pikake?" Marco strained.

"Marco!" Mrs. Sanders stood from her desk. "Thank you so much for meeting with us today." She moved to the middle of the room. Mr. Pikake turned but stayed like a statue at the window. Marco kept his eyes on his tutor. "Your tutor has just introduced himself to me. I had to meet the genius who has been doing such a wonderful job with my student!" Mr. Pikake gave the teacher a kind smile although Marco could feel the tension in the room. The way Mrs. Sanders said *my student* implied Marco was her property, not Mr. Pikake's.

After a nervous chuckle she continued, "I was hoping he'd share some of his secrets!"

As if sensing Marco's nervousness, Mr. Pikake swooped in to the rescue, "Hello, son. Mrs. Sanders just wanted to check in on what we are working on. I shared that we had just finished solving one-step equations and inequalities with addition and multiplication, and are beginning to work on more complex problems." With two blithe steps, he had crossed the room and was at Marco's side. Placing his hand on Marco's shoulder, he gave the boy a strong squeeze.

Marco was spellbound. He didn't understand what was happening. Mr. Pikake *never* talked about math and numbers in plain teacher-speak. His tutor was a master, a commander, a vanquisher, an obliterator. He was not a boring *teacher*. He sensed that Mr. Pikake was wearing his own mask. He didn't want Mrs. Sanders allowed into their special world.

"*Um*...Yes...I can show you if you'd like?" Marco squeaked out.

$$\frac{x}{17} \geq 17$$

"I would just LOVE that, Marco!" his teacher yelped clasping her hands together.

Moving to the chalkboard, Mrs. Sanders wrote down her scenario:

$$\frac{-1}{16}\left(x - \frac{7}{3}x\right) \geq 2.$$

What the heck is this? Marco thought. Something about being in his math classroom made him feel small, like all his power had somehow been sucked away. It didn't help that Mrs. Sanders threw an ugly situation at him either. He looked over his shoulder, where both adults stood anxiously waiting for him to continue. *And I have an audience.* He forced a quick smile before looking back at the board. He needed a mathy explanation, one that wouldn't alert his teacher of his training with the SAN. Knowing he would need to choose his words carefully, he surveyed the scene.

Okay Marco, he started, *what are you going to do?* The inequality came to life as an alligator burst out of the chalkboard. At first Marco welcomed the reminder, he'd need to keep that in mind as he battled. It was when the alligator stopped going for the expression and started snapping *at* Marco that his rollercoaster rushed forward at lightspeed. He felt dizzy, he might faint.

Do what you can, start with what's easy. Convincing himself he could do this, a new image appeared on the board. Werewolves. As the *x* transformed into the snarling creatures, they barked at the alligator. The reptile snapped its long snout in return. *Like terms! Simplify the house first, then worry about the suppression spell.* Calming down, he chuckled at the $\frac{7}{3}$ werewolf. He forced it to be two werewolves and another third. Drool dripped from the partial mask that was made only of a tiny part of the snout and about half of one eye.

Marco remembered that when Logos dueled, they had to have a common denominator. Mr. Pikake shared that these battles were like circus sideshows – Integers watching in awe as the Logos

$$5x \leq 1450$$

transformed themselves. This helped. The werewolves and alligators faded away as the parentheses became a big top circus tent. Inside, the $\frac{7}{3}x$ was a clown now. The 3 balanced a long bar on its head as the 7 attempted to stay upright wavering back and forth atop the bar. *Much better*, Marco thought as the coaster came to a crawl.

"First, I will combine the like terms," Marco announced to the room. "The x is really $1x$," *because ones have the power of invisibility.* He was careful not to say this out loud. "And 1 can look like $\frac{3}{3}$." He wrote:

$$-\frac{1}{16}\left(\frac{3}{3}x - \frac{7}{3}x\right) \geq 2.$$

Duel! Duel! Was what he thought, but what he said was, "Now three minus seven is..." The 7 clown launched a confetii cannon and out came the seven minuses, then 3 did the same revealing three pluses. The three pluses and three minuses combined, sending sparks shooting, as the four remaining minuses drifted slowly to the ground, "...is negative four."

$$-\frac{1}{16}\left(\frac{-4}{3}x\right) \geq 2.$$

Now what? I can obliterate with a -16, but... Marco didn't much like that idea. It would force him to deal with the alligator before he was ready. As he stared, the four began hopping back and forth from the top of one Vinculum to the other. *Multiplication. The suppression spell is impacting everyone in the tent.* Allowing the four to stay on top of the sixteen, Marco silently commanded him, *take the negative with you!* Before continuing...

"I, *um*, want to clear things up a bit first, so..."

$$-\frac{-4}{16}\left(\frac{1}{3}x\right) \geq 2.$$

$$\frac{x}{2} > 145$$

"Now, the negatives make a positive and four sixteenths is really one fourth."

$$\frac{1}{4}\left(\frac{1}{3}x\right) \geq 2.$$

As he wrote, the chalk made a loud squeak that broke Marco's focus.

"The suppression spell turns the left to a twelfth of x."

$$\frac{x}{12} \geq 2.$$

And obliterate it! "Now I can," Marco paused, he wanted this to sound as mathy as possible to leave no doubt in his teacher's mind, "multiply by the reciprocal of the coefficient to find,"

$$(12)\frac{x}{12} \geq (12)2$$
$$x \geq 24.$$

Perfect! That was an amazing demonstration of my skills, Marco thought. He turned, standing tall and expecting to see both teacher and tutor beaming. In his mind, they were wearing huge smiles and ready to shower the boy with praise. In reality, what he saw was quite different.

Mrs. Sanders had a blank stare on her face, her jaw hanging slightly open. Mr. Pikake looked like he had eaten something sour – his face twisted in a mix of anger and shock. It was like Marco had both forgotten to distribute over the parentheses *and* figured out how to achieve world peace.

"Ahem." Mr. Pikake was the first to speak, starting with a theatrical throat clear. "Kids today! Everything's magic and spells. Must be all that *Dungeons and Dragons.*" He pushed out a laugh.

Taking the professor's cue, Mrs. Sanders laughed along, "Oh yes! Yes! I often make my word problems about video games to relate." She turned her attention to Marco, "What a beautiful job you have done! I have to admit, you lost me with talks of suppression spells, but whatever works!"

Marco must have gone through six shades of red realizing what he had done. He shot Mr. Pikake an 'I'm sorry' look, but the tutor only turned away.

"I have to admit, your method even got around my little trick. Most students would solve," she took the chalk and quickly pecked at the board:

$$\frac{-1}{16}\left(x - \frac{7}{3}x\right) \geq 2$$

$$-\frac{1}{16}\left(-\frac{4}{3}x\right) \geq 2$$

$$-16\left(-\frac{1}{16}\left(-\frac{4}{3}x\right)\right) \leq -16(2)$$

$$-\frac{4}{3}x \leq -32$$

$$-\frac{3}{4}\left(-\frac{4}{3}x\right) \geq -\frac{3}{4}(-32)$$

$$x \geq 24.$$

"One of my trickier problems to make sure students remember to flip the sign when multiplying by a negative, but you managed to avoid that entirely." She tilted her head to the side examining her own work, "I am going to need to rethink this one..." she trailed off before spinning around merrily. Marco had forgotten all about flipping the inequality sign. He felt lucky to have avoided that mess. Mrs. Sanders continued, "Well, I am so happy things are going well. I have been so impressed with Marco's growth this year. He has become a top student in class." Glancing at the clock she began talking more quickly, "I have to run to a department meeting. I want to thank you both for your time. Great job again, Marco!" With that, she exited the room.

Like a wounded puppy Marco slowly approached Mr. Pikake. He was back at the window, his head down. "I'm so sorry!" Marco exclaimed, "I didn't mean to say it aloud. I know you have only

$$4x - 21 \leq 1151$$

told me about the Numberfolk because you trust me. I really didn't mean to tell."

After a long pause, Mr. Pikake turned. Marco was shocked to see it wasn't anger on his face at all, it was... sadness.

"I think it is time I give you the full story, Marco," he said somberly. "I, too, wish we didn't have to keep secrets. I believe the whole world should know. But people, people forget their own history, they like to push it away and cover it up. They've twisted the story so they can sleep at night. And what's worse is this twisted tale has become so common, people wouldn't believe the truth even if they knew."

The tutor led Marco to a desk. Marco slid easily in the small space between the chair and the tabletop. Mr. Pikake took a seat in the neighboring chair, his long frame stuck out of both ends like a giant on a baby toy.

"When I was a young child, I was excellent with numbers," the professor began. "I learned much like you have. My schoolteachers would tell me what to do, what steps to follow, what tables to memorize. Despite that, I had grown to truly love numbers. I was skilled at manipulating them and using them for my needs. While the rest of my class counted on their fingers to solve something as simple as $8 + 7$, I rearranged the problem to read $8 + 2 + 5$ to quickly shout out 15 before they even counted to ten. Every math problem was a new puzzle, a new game, and I loved games."

Mr. Pikake turned from Marco to stare out the window. "One day, my father took me into his study and told me all about the history of Numberville. He told me about the wars, about dueling evil twins, about the Vinculum Games. And this made me love numbers even more. They were like people to me. They were friends who never left my side." He let out a strong sigh, "My father saw how much numbers meant to me and so he told me a new story. He told me about how numbers control the world."

$$\frac{x}{2} + 27 \geq 174$$

Marco felt like he knew this already. He remembered the giant 15 jumping down and attacking him from the clock, the giggling 5's, and the temperature 2's. That was why he had become so interested in working with Mr. Pikake. Marco had started to see the numbers all around him and wanted to be able to master them, to not be afraid of them.

Mr. Pikake turned back to Marco and looked him straight in the eyes like a parishioner ready to list their sins to their priest. "I haven't told you the whole story, Marco. For many reasons. Because so few can truly see it. Because it is frightening. Because I believe in you." Pop. Pop. Pop. Pop. "Let me give you an example. Say you are up to bat at a baseball game. The pitcher throws the ball. You swing. You hit! You start running the bases. How much do you think numbers interfered?"

Marco imagined the scene. He got ready and steadied his feet. He kept his gaze firm on the pitcher's glove. The ball came flying toward him. He hit it with incredible force and began running the bases. As he did, he looked around, there was not a number in sight. He relished in the moment, letting the adoration of the cheering fans take him over, until, out of the corner of his eye he saw the number 5. It was giggling on the back of the second baseman's uniform. This time, it wasn't a fun giggle, it was sneering at Marco, taunting him.

He turned to see the scoreboard – more numbers and they were snickering at him, too. Marco tripped on second base. His entire body came to a sharp thud as his face hit the ground. Not wanting to continue in his made-up world, Marco returned to Mr. Pikake. "There are numbers on all the jerseys and on the scoreboard, too!" he shared.

Mr. Pikake bowed his head. "Yes Marco, but those are only the numbers you *see*. Those are not the numbers in control. As we sit here in this room, there is a 32 pushing down on us, it's the 32 that determines gravity. There are also numbers all over the floor. Over here," he pointed to a rug near the window, "well, that has

1.3's hanging around it." He pointed to the tile, "Over there, that has 0.42's that will push against you when you walk across it." Mr. Pikake stood and began to walk away from Marco. "We talked before about the weather, the temperature, the air pressure, the UV index – all Numberfolk. The arc of the ball when it's thrown, the force of the pitch, more Numberfolk. The tilt of the planet, and everything that is keeping this building from crushing us alive – all Numberfolk."

He turned back to Marco. "We work to thwart the Numberfolk we can't control every day and attempt to control those we can." He pointed to the thermostat, "That was created to keep us safe. While we can't control the many variables that dictate the weather, there is a number we can control to make sure it is warm enough or cool enough inside for human life. All of history is riddled with inventions and breakthroughs to control what we can!" He threw his arms in the air. "And there is story after story of the times we failed. The times, despite the best effort of humans, the Numberfolk won. A building in Las Vegas allowed Numberfolk to attack guests through reflections and light waves from a building constructed as a parabolic mirror. The Numberfolk are certainly celebrating over that one. They too laugh every time they even think about the Wobbly Bridge in London."

Mr. Pikake made long quick strides and was back at Marco, he knelt in front of him. "But the real problem is that people, everyday people don't understand. They don't understand the Numberfolk are everywhere and worse, they don't understand the numbers! A campaign of people believed a program that cost $360 million for 317 million people was the same as giving everyone a million dollars*. We are losing the war Marco… we are losing it every day. As less and less people understand, the Numberfolk gain more and more ground. We brand math as too hard for some.

---

* Can you do the math? Such a program is about $1 per person not $1 million per person!

$$-\frac{1}{8}(x+4) \leq -\frac{75}{2}$$

We place labels saying there are children who are simply not capable of controlling the Numberfolk, and as a result, the next generations will be doomed."

Jumping back to his feet, his voice got louder, words came more quickly. "And we haven't even discussed the numbers *inside* humans. They regulate our hormones, our blood pressure our heart rate. We do what we can to keep them in control but more than a few know how hard that is. Those numbers take over and we are done for."

Marco was back on the baseball field. He raised his head to see the crowd. They were still yelling and cheering but not *for* Marco. They were yelling *at* Marco. Like the numbers, they were taunting him. He pushed himself up, he wasn't sure he was still on the field. Everything looked the same – but different. Numbers were everywhere. He saw $45°$ and $763.6875\sqrt{2}$. He saw arrows in all directions and each one had a number attached, 9.8's and 0.49's. He started to run.

As he took off, the arrow on his chest grew larger…it was sticking straight ahead of him now. It was 15 – now 16! He came to a sharp halt. The arrow retracted back to his chest displaying a zero. He felt a gust of wind and looked up to see numbers riding the breeze. He saw angle markings and arcs. An outfielder threw the ball towards third base and with it came numbers and letters in strange forms Marco didn't even recognize. He closed his eyes and shook his head hard. It wasn't until he felt Mr. Pikake's strong grip on his shoulder that he opened them again to find himself safe in the classroom.

"I-I" Marco struggled to speak. He could hear his heart pounding. Looking down, he was relieved to see the arrow was no longer there. He blinked and a 98 appeared pulsing in and out of his chest. Now it was a 99, now a 100.

"Breathe Marco, breathe!" He gave the boy a light shake. "You see it now, don't you?"

$$-2(x - 50) \geq -494$$

Slowly inhaling then exhaling, the beating value on Marco's chest began to decrease, it was 80, now 77. Marco squeezed his eyes tightly. When he felt he was regaining control, he opened them thankful to see the pulsating numbers had vanished. With a deep sigh of relief Marco tried to talk, "W-w-what happened?" Tears began to pool in the student's eyes.

"You saw them, Marco. You saw the numbers. They control everything around us and within us. Long ago when humans began to realize this, out of fear they established the SAN – the Society for the Abolishment of Numbers. My father was a member, as was his mother, and long down our family line. But the society is dying, the appreciation of mathematics and its importance is dying, and, well, I am dying, Marco. With no children of my own, I have been searching for a long time for someone to pass down the knowledge and the power of the SAN to. And, well, I'd like that someone to be you."

Marco realized he had made a grave mistake. When Mr. Pikake talked of Numberfolk, Marco focused on the simple ones. The ones that willfully made themselves known or could be easily detected. The numbers on a clock, the temperature outside, how charged his phone was, speed limits, addresses, the number of steps he took in a day. But these were not the numbers his tutor feared, his tutor feared the unknown numbers. The ones that are hiding all around us, hiding inside us. The numbers that controlled our bodies or the powerful forces of gravity and friction that surrounded us at all times. What was worse was that these numbers worked together in ways Marco was far from understanding. They possessed a trained and diligent army. Marco, well he wasn't even a math kid.

His blood started to boil, "My teacher told my parents I wasn't a 'math kid'! What was she trying to do? Make me a defenseless pawn?" Marco let the anger take over. Emboldened by Mr. Pikake's belief in him, he decided at that moment he would do whatever it took to become a master. He looked his tutor in the eyes, "What do I need to do?"

$$500 - 3(x - 6) < -373$$

* * *

"I assume everything is going to plan with *this* one?"

Mr. Pikake nodded. "He is making great progress. This is the one, I know it."

"You had better hope so. I don't need to remind you of the mess we were left to clean up from your other trials. Nor of the consequences should you not identify an heir."

"I understand. I understand."

HUNTING 101 UNIT 3

$$\frac{1}{25}(x + 625) < 37$$

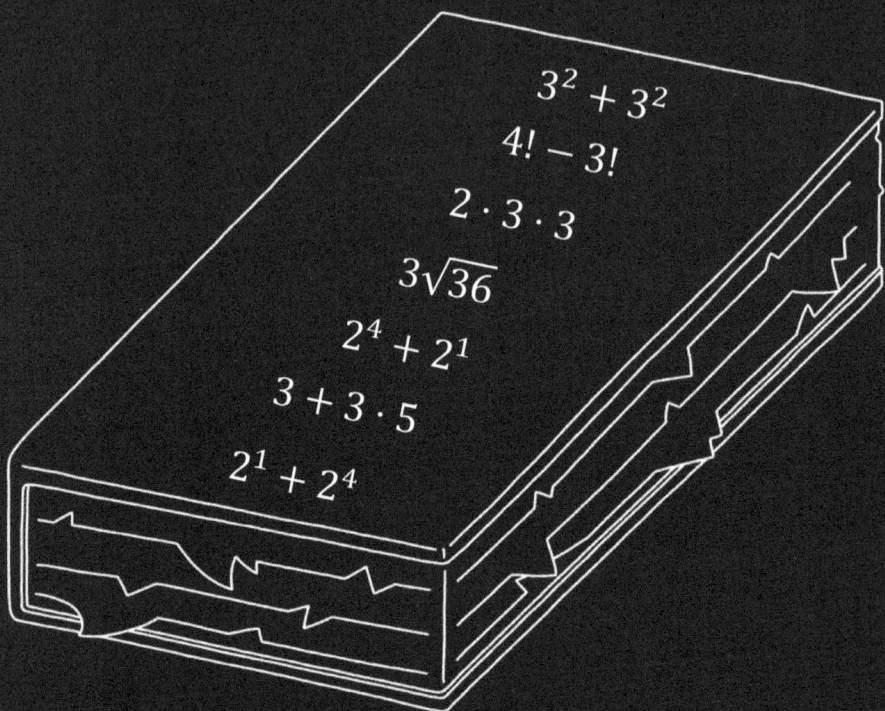

$3^2 + 3^2$

$4! - 3!$

$2 \cdot 3 \cdot 3$

$3\sqrt{36}$

$2^4 + 2^1$

$3 + 3 \cdot 5$

$2^1 + 2^4$

18

A DREAM AND A NIGHTMARE

"FOR EVERY ACTION THERE IS AN EQUAL AND OPPOSITE REACTION."

-SIR ISAAC NEWTON

The week flew by. Mr. Pikake stressed to Marco the dangers of sharing what he knew about the SAN. They were a secret organization and would do anything to protect themselves – including harming his family and friends.

Marco had no choice but to distance himself from everyone. It saddened him that this was easier than he expected with Oliver, Peter, and his mother already far-off shadows in his life. Marco followed his schedule: school, baseball, dinner, homework. Each night he worked hard practicing his level 1 hunting, mastering both vanquishing and obliteration.

Staring down at his bowl of cereal, silent, he wasn't even paying attention when his mother started jabbering on.

"Marco? Marco!" she yelled to get his attention.

"Huh, oh, yeah?" he responded without looking up.

"Did you hear me? Today is your first game! Maggie and I will be there in the front row cheering you on!"

Marco had almost entirely forgotten. It was strange for him. He had been looking forward to the start of the season for weeks, now it somehow seemed, unimportant.

"Thanks Mom," he said swishing the little circles around in the milk. They looked back at him with eyes shaped like π and an r for a nose.

School came and went. Before he knew it, Marco found himself suited up completing his pre-game drills on the field. His mother and sister were already happily seated in the front row of the stands as promised. Maggie waived fanatically and Marco gave her a quick raise of his hand before tossing the ball to second.

Numbers were everywhere. Marco couldn't escape them. They were a stain on his vision. It was like the birthmark in his eye. Normally invisible but once he noticed it, it suddenly appeared everywhere he looked. The velocity of the ball, the speed of the

$$5(x - 11) = 3(183 + x)$$

wind, the pull of gravity, the length between bases, the curvature of a pitch, everything was numbers.

The game began and Marco tanked. Between the numbers and the sun, he missed a ball and an out. Luckily, his teammates were doing well and play switched. Unfortunately, he easily struck out as the combination of the ball's path and speed made it difficult to even know where to swing. It used to be a feeling, now it seemed like an equation he needed to conquer. Embarrassment took over.

"Marco, don't worry. It's the first game. We all need to get the kinks out." Oliver encouraged him from inside the dugout. It was an olive branch, but Marco was too upset to take it. He ignored his friend and buried his face in his hands. He couldn't look into the stands, he didn't want to see his mother and sister's faces flush with disappointment.

He was sent back out to first base. As he jogged to his position, he saw something out of the corner of his eye. *It couldn't be? Could it?* His curiosity won over his fear as he glared into the bleachers. His mother and sister were still cheering enthusiastically despite his subpar performance. But that wasn't what interested him. He tilted his head (37° to be exact) to look at the very last row in the far-left corner. There Mr. Pikake sat.

A wave burst through Marco. His roller-coaster took him up, down, sideways, and upside down. He thought he was going to burst into tears. No one had ever come to his games except his mother and sister. No one cared enough about Marco to support him, to cheer him on. He pushed back the tears, then he pushed back the numbers, then, well then, he played ball.

He was fantastic! He jumped and he dove, he threw, and he hit. After each play, he'd look and see his tutor cheering him on.

❊ ❊ ❊

By Saturday, Marco was still floating on a cloud from his baseball win. Word had gotten around school about his impressive come back. All-day Friday, people he didn't even know were

$$6x - 2(x - 8) = 925 + x$$

giving him high-fives in the hallways and congratulations in the forms of 'way to go Marco!' shouted from across the room.

As Liam and Marco were leaving school late Friday afternoon, he started in. "So Marco. I was thinking…your tutor *has* to be part of the Illuminati. All the signs are there. It screams secret society more than I've ever seen before."

Panic struck Marco right in the throat. He didn't know what to do. Liam needed to stop digging. With no other ideas, Marco went off. "Dude, you have to stop with all this nonsense. Do you know how stupid you sound?" He cringed, he sounded just like Peter. "We've all been too nice to say anything, but it just can't go on. It was alright when we were younger, but we'll be in high school in no time. You'll get pulverized talking crazy like this all the time."

Liam was stunned. With a look like he was about to cry, he pushed Marco to the side and ran down the hall. Marco thought he was going to throw up. It felt horrible. But he had to do it. He couldn't have Liam snooping around and getting himself into trouble.

Marco lay on his bed tossing the ball into the air. Having convinced himself that he actually helped Liam, he only allowed the good feelings from his game to penetrate his thoughts that Saturday morning. Without warning, his door flung open. He turned his head to see his sister storming in.

"Big star, huh?" she said sarcastically jumping onto the bed. "Too good for me now? I feel like you haven't talked to me in *days*. What is going on with you?"

Marco groaned and rolled over. This was just the thing to bring down his mood. Out of the corner of his eye, he saw his journal. Maggie must have seen it too as she dove to grab it. The two fought over the pad of paper.

"Just let me see it!" she shrieked. "What is your problem? Are you hiding something? Does this have something to do with why everything in the bathroom has tape on it?!"

$$\frac{1}{8}(x + 128) = 54$$

Then it happened. With all his might Marco shoved his sister away. One hand gripped tight to the journal, the other firm on her shoulder. Before he even realized what he had done, she was flying off the bed. As if in slow motion he saw her tumble down, her hand shooting out to catch herself. Numbers whisked everywhere. Her velocity, the acceleration from gravity making her move faster, pulling her towards the ground harder. Then she hit.

He wasn't sure what came first, the crack or the scream. Both were deafening. Tears streamed down his sister's face, but no other sound came out. He had watched as her wrist caught the full force of her weight and the fall, jutting out at an angle he had never seen before – at an angle that didn't seem possible. Marco only had time to blink before his mother and Peter were at his door.

"What did you do to *my* daughter?!" Peter roared leaping across the room and grabbing Marco's arm so hard he thought it would detach from his body. He looked to his mother, whose face was frozen in shock. She finally saw it. Peter had made the grave mistake to show his monster-mask to someone other than Marco. Unfortunately, this time Marco deserved it – he had beckoned the monster.

Marco couldn't move. He couldn't speak. What would he even say? If Maggie started digging into his notebook, she'd quickly begin investigating the SAN. He knew how she was. A dog with a bone, she wouldn't give up, she'd put herself in danger. But he couldn't justify that to his mother.

Peter and Maryanne scooped Maggie up and ran out of the house without another word. By noon they had returned from the hospital. Maggie had a bright pink wrist cast...the fall on her hand had snapped a bone. Their mother ushered Maggie to her room and neither of them even looked at Marco.

"Get in your room," Peter barked pushing Marco through the doorway. Once they were alone, he bent over to meet Marco's eyes and in an unnervingly clam voice continued, "One thing I

$$9x - 61 = 4(x + 366)$$

know is that the best way to deal with something you can't kill is to lock it up in a cage. Don't even think of leaving this room." Marco knew better than to respond. He deserved much worse. He hated himself.

Not even an hour later his mother quietly pushed his door open, with her head down she whispered, "It's time for your tutoring." Marco was stunned. He was sure he would be locked in this room forever – until he decayed into dust. Wanting nothing more than to flee his house, where everyone looked at him like he was the monster, not Peter, he quickly packed his things and raced out.

He explained the entire event to Mr. Pikake. He was the only one who could begin to understand.

"Don't be so hard on yourself." His tutor squeezed his arm. "You were doing good, you were trying to protect her." While it was a nice idea, it didn't convince Marco and it certainly didn't make him feel any better. They sat in silence for what felt like forever. "Would you like to take the day off?" Mr. Pikake asked gently.

The last place Marco wanted to be was home. He could feel the anger and disappointment in every room. Plus, this had to be worth it. The secrets, Maggie's wrist. This all couldn't have happened for Marco to fizzle out and become a failure. He wanted to be a master, he needed it.

"No. I'm ready," he said sternly.

"Our content for today is challenging. We now begin to explore more complicated camouflages. Level 2 hunting as it is known."

"Bring it on."

Respecting the student's wishes, Mr. Pikake began gently, "Let us commence with when we can vanquish."

Marco nodded and stoically responded, "We can vanquish when there is a duel."

"Very good. And how do you know if there is a duel?"

$$\frac{1}{3}(2x + 9) = 309 - \frac{x}{3}$$

"If there are terms being added or subtracted." Marco was starting to relax. He appreciated the soft balls the tutor was pitching.

"Perfect. And how do we vanquish?"

"We force the numbermask to duel with Zero. We do that by sending in the evil twin to duel with both sides of the scales."

"Excellent, son! And when do we obliterate?"

"We obliterate when a numbermask is hiding in a group. The numbermask is either disguising itself through enlargement as three c's or $3c$ or has split itself into parts, suppressing the mask to hide like $\frac{3}{4}m$."

"You are doing amazing, lad! And how do we obliterate?"

"Obliteration is sending in the reciprocal to combine, er, multiply with the coefficient."

"Superb! Now, the situations we explore today will require both vanquishing and obliteration. These become very tricky because the Numberfolk we see, well they aren't always the Numberfolk who are actually involved. Remember, they are disguising themselves! You must move deliberately and carefully consider the situation."

Marco was ready. He imagined himself a wizard going into battle. He had two main spells, vanquishing and obliterating, and he was ready to use both.

"Let's take ourselves back in time, to when we first met." Mr. Pikake's words came out as one long string, not the precise pronunciation Marco was used to. Like a lullaby, they were calming, comforting. Marco struggled to remember life back then. It wasn't even that long ago though it felt like a lifetime. The professor scribbled:

$$\frac{9}{5} \times C + 32.$$

$$74 - 2x = 4(172 - x)$$

"Do you recall this?"

The familiar expression that allowed Marco to change Celsius to Fahrenheit was child's play. No vanquishing or obliteration, a simple act of substitution was all that was needed. This couldn't be the complex situation his tutor was referring to. "I do." Marco responded, his words lacking any emotion.

"Wonderful!" Mr. Pikake attempted to lighten the tension with his voice. "We could say that this expression is equal to F, Fahrenheit."

$$F = \frac{9}{5}C + 32.$$

"My question for you is this. Can you twist this into an equation for C? Manipulate, master, make known."

"Are you asking me to hunt C?"

"Exactly! Begin by vanquishing the 32."

"Okay. Well, I'll send in 32's twin to duel." His words lacked energy, they lacked life.

"Splendid. Endure," the professor said nudging his head forward. Marco began to drag the marker across the board. It wasn't the wizardly slashes of Mr. Pikake, nor was it the avian pecking of Mrs. Sanders.

$$F + (-32) = \frac{9}{5}C + 32 + (-32)$$
$$F + (-32) = \frac{9}{5}C.$$

"Yes, yes…" Mr. Pikake's eyes grew wide. "Now obliterate!"

"Well, so I cast the $\frac{5}{9}$ to both sides." The power of the hunt was starting to leak into Marco. His command of the Numberfolk was helping him to regain his strength.

$$\frac{x}{4} - 7 = 70$$

$$\frac{5}{9}\left(F + (-32)\right) = \frac{5}{9}\left(\frac{9}{5}C\right).$$

"Very good! When you cast your spell, it hit everyone. Not only the F or the -32."

"The right is obliterated so," Marco was now slashing, not at the speed or precision of the professor, but certainly increasing his momentum.

$$\frac{5}{9}\left(F + (-32)\right) = C.$$

"Perfection!" Mr. Pikake jumped in.

"Don't I need to apply the suppression spell? Distribute it to everyone inside?"

"You could, but it is not required. Do you remember when you asked me how one can change a Fahrenheit Numberfolk to a Celsius one?" He turned. His arm floated through the air like a game show host presenting the contestants. It stopped at the equation on the board – Marco's equation.

"You mean... you mean this is the formula? *I* created a formula?" Marco felt the blood rushing to his cheeks, "Er. f – word??"

The Cheshire cat smile hijacked Mr. Pikake's face, "You did that, son. I hate that word, but you did just that. Now, it is 86 degrees Fahrenheit, what is it in Celsius?"

That's easy. Marco began slashing. He was quick, methodical, fluid, and smooth. He never hesitated or slowed down.

$$\frac{5}{9}(86 + (-32))$$
$$= \frac{5}{9}(54)$$
$$= 5\left(\frac{54}{9}\right)$$
$$= 5(6)$$

$$2(x - 200) = 527 - x$$

$$= 30.$$

Marco remembered. He remembered the beginning of their story together. He remembered changing the thirty degrees Celsius to the balmy eighty-six degrees Fahrenheit. And look at him now! He was the apprentice and Mr. Pikake, the sorcerer. And just like Mr. Pikake promised, Marco didn't need to memorize this. He *created* this! It was *his* formula now, not a stupid rule nailed to the wall and etched into his brain. He felt liberated, he felt mighty, he felt happy. He started laughing, he laughed loudly, he laughed hard. For a nanosecond guilt struck him. How could he be so happy when Maggie was at home feeling so badly? He pushed it away before the thought could even fully form. *This is for Maggie. I am strong now. I can protect her.*

"A formidable force you are!" The professor shrieked in delight.

Power is a funny beast. A curse. Once you have a taste, you can never be full. As it soaked into the boy, he didn't find satisfaction, he felt famished. He wanted more. Like a rabid dog he looked to his leader and growled, "Keep going."

Mr. Pikake threw the canine a bone:

$$4n + 6 = 26.$$

Marco understood the task. More ready than ever to turn Numberfolk into zeros and ones, to blast them from existence, to hunt them down and lock them up, he considered what he must do. He needed to vanquish the 6 by forcing a duel with its evil twin and he needed to obliterate the 4. What he didn't know was how to go about it. His rollercoaster felt wobbly, like the smooth metal tracks had turned to wood. He realized he had gotten ahead of himself. He still had a lot to learn. What he knew for sure was that he would learn, he would learn for Maggie.

"What should I do first?" he asked.

"That is the question. Unfortunately, the answer is not so clear. Sometimes you vanquish first, other times you obliterate first. It

$$\frac{9}{10}x = 31 + \frac{4}{5}x$$

all depends on the situation – the battle at hand. In most cases you can do either but," pop, "the disguises are tricky. They can lead you astray."

This was not what Marco wanted to hear. He hated rules, but he had to admit it was nice to have a clear battle plan going in. Having to decide on the fly made him feel exposed, vulnerable.

Marco glared at the situation, he saw the $4n$ and the 6 as knights, the plus sign became two swords crossed in battle. With the obvious duel taking place, he decided to vanquish first.

"I am going to send in the evil twin of 6 to duel."

$$4n + 6 + (-6) = 26 + (-6)$$
$$4n + 0 = 20.$$

"Excellent, excellent." The tension was dissipating.

"Now $4n$ duels with 0 to result in $4n$."

$$4n = 20.$$

"What shall you do now?"

"Next, I obliterate. The reciprocal of 4 is $\frac{1}{4}$, so I will slash both sides by 4."

$$\frac{4n}{4} = \frac{20}{4}.$$

He continued, "The left is obliterated. So, we have a single n standing alone. The right breaks 20 into 4 parts. So, we have…"

$$n = 5.$$

"Beautiful, son!" the professor cheered. "What if you had obliterated first? What would be the result?"

As Marco thought about the question, he wasn't sure of the answer. He was starting to feel weak again and he didn't like it. He didn't know how to obliterate when there were duels going on. All the work before had been only enlargement or suppression

$$8(x - 50) = 4(211 + x)$$

spells, no duels. "I guess I would do the same thing. To obliterate the 4, I have to send in $\frac{1}{4}$."

"That you do!" The professor wrote the hunt out again.

$$4n + 6 = 26.$$

"The problem is that when you obliterate, when you send in a suppression or enlargement spell, you hit everything in sight!" Mr. Pikake continued.

Marco remembered this idea – it was the same as parentheses. When the Numberfolk were working together, they were grouped which meant any spell would impact them all. Guessing, he wrote:

$$\frac{1}{4}(4n + 6) = \frac{1}{4}(26).$$

"Like this?" Marco asked.

"Exactly! When you obliterate, you cannot pick and choose who your spell will hit, no matter what, it impacts all, it hits everyone on the right and everyone on the left."

The problem was suddenly much more complicated. Marco felt the sweat slowly creeping down his back. He might as well figure out the effects of his suppression spell to get a better idea of what he was dealing with. He distributed.

$$\left(\frac{1}{4}\right)4n + \left(\frac{1}{4}\right)6 = \frac{26}{4}.$$

"Perfection!" The professor squeaked. "Now what?"

"Well, the one-fourth obliterates the 4 like before, so I get…"

$$n + \frac{6}{4} = \frac{26}{4}.$$

"I suppose now I duel, since that's all that is left?" Marco looked to Mr. Pikake who nodded him ahead. "Alright. So n is duelling with $\frac{6}{4}$ so I send in the evil twin, $-\frac{6}{4}$, to duel both sides."

$$\frac{3x}{4} = 312 - \frac{1}{4}x$$

$$n + \frac{6}{4} + \left(-\frac{6}{4}\right) = \frac{26}{4} + \left(-\frac{6}{4}\right).$$

Marco continued. "The 6/4 is vanquished by its twin on the left and on the right 26/4 minus 6/4 is 20/4."

$$n + 0 = \frac{20}{4}.$$

"So 20/4 is hiding behind n? But that isn't the same as I found before." Marco knew this was going to be more complex, but he didn't expect to feel so angry about it. He took a few deep breaths trying to regain his control.

"*Ahh!* Yes and no. It is true that 20/4 is hiding behind n, but 20/4 is yet another mask! Who is 20/4 really?"

Then Marco saw it, he had forgotten to complete the final step. The 20/4 was an Integer disguising themselves, pretending to be a Logos they weren't! The mask of twenty divided by four... Marco shouted, "Oh! It's five! It's five both ways! I see it. The second hunt was way harder, let's not do that again." He was starting to feel like himself. It was like a suppression spell had been cast on Marco making him small and the professor had maintained his head and obliterated it, bringing the student back to his full form.

Mr. Pikake let out a loud chuckle. "Right on, son! Right on! Many will make the mistake of casting the obliteration but thinking it only effects the $4n$ leading to the wrong Numberfolk trickster. Remembering that your spell impacts everyone is the key." He presented Marco with a new scenario.

$$18 = 3(x - 5).$$

Marco worked out what was happening. He saw the $x - 5$ huddled together inside their home. An evil Visionary witch cast a spell on the house making its inhabitants three times bigger. But what should he do? Should he vanquish the -5 first or obliterate the spell?

$$3(x - 214) = 923 - 2x$$

Still deliberating, Marco was snapped back to reality by Mr. Pikake's voice, "A penny for your thoughts?"

"I am not sure if I should vanquish the −5 first or obliterate the 3."

"Yes, yes. Clearly the conundrum. May I ask you…I know we see −5, but is that really who is present?"

He instantly understood the tutor's hint. The evil witch's spell had been cast. What once was a −5 was no longer. What was really there was a −5 who had been enlarged three times its size. "I see, it isn't really −5 because the 3 is acting on the house."

"Precisely. Perhaps you may find this helpful. You have many ways you could complete your hunt. Let us boil them down to two. The first is to unobstructify. If you do this, the field is clear – you shall know exactly what is transpiring. Then, you will always vanquish and finally obliterate."

Marco tried to commit the tutor's words to memory, *unobstructify, vanquish, obliterate.*

"The second selection is to hunt as is. To go in headfirst and begin slashing away at the Numberfolk who are aiding and abetting the mask. This means that in some cases you will obliterate first and in others, you will vanquish first. This method is for the master who can clearly see the battlefield for what it truly is through all the illusions: who is helping, who is deceiving. Let us tackle both trickeries so you can debate which you desire?"

Marco nodded, determined to prevail. "Okay. So, the first method is to: Unobstructify. Vanquish. Obliterate. How do I unobstructify?"

"Numberfolk are intricate illusionists. In the case of $3(x − 5)$, we know the enlargement spell has been cast on the house, but we don't *see* who is really there. To unobstructify you must finish the enlargement and suppressions that have been cast, reveal who is

$$\frac{x}{4} - 1 = 156 - \frac{x}{4}$$

truly there, enact any duels you can. We call this the Recta Defensive."

"The Recta Defensive?" Marco leaned in. Now they were getting to the good stuff.

"Numberfolk disguise themselves in many ways. Sometimes their trail is tidy, straight. Other times they are curved, twisted, and turned."

"Like the path of a baseball! That's curved!" Marco jumped in.

"Exactly! Other times, they are waves." The professor slid his hand through the air moving it up and down again and again. It remined Marco of his roller-coaster, up, down, back up again. *But what actually acts like that?* Not able to think of anything, he wanted to know. "Like what?"

The professor hesitated. He didn't want to completely terrify the young boy, so he treaded carefully. "Everything we see, everything we hear. We accept that what we see is really what it is. When I look at this marker," he pointed to the board, "I see blue. However, blue is simply the result of light waves hitting the object. And light waves, like sound waves, are controlled by Numberfolk."

Panic clenched Marco's chest. *They even control everything we see and hear? That means they control my entire reality.* He was transported to his computer. Comfortably seated in the chair satisfied he had beaten his sister to the office, suddenly he was sucked into the game. He turned to see the evil 15 at the controls. With every click, Marco was forced to do as the number demanded. He evicted the thought from his mind. Feeling nauseous, he pushed himself to continue. *Gulp.* "Okay, let's save that for another day. What is the Recta Defensive?"

"Ah yes! We are beginning our hunting only worrying about those disguising themselves by taking a straightforward path. And anyone taking a straight path can be boiled down to $ax + b$." He scribbled the expression down. Marco's own rollercoaster was

$$x - 10 = 200 + \frac{x}{3}$$

bouncing up and down, up and down. Having just barely gotten over the last shocker, now he was presented with an expression with no numbers at all! "Don't worry. This just means x is the numbermask you are hunting and, a and b are whomever is helping it. Try it out." He motioned to the hunt that lay between them:

$$18 = 3(x - 5).$$

"Okay. I unobstructify, vanquish, obliterate. First, the 3 is enlarging everyone in the house, so I should figure out who they really are."

$$18 = 3x - 15$$
$$18 = 3x + (-15).$$

"Very nice! You see son, the $ax + b$ is simply that a is 3, and the b is the -15. You have set this up perfectly to complete the Recta Defensive! You may conclude your hunt by vanquishing then obliterating!"

This was now easy work for Marco. The $3x$ and the -15 had their swords drawn, Marco forced the duel with -15's evil twin, 15.

$$18 + 15 = 3x + (-15) + 15$$
$$33 = 3x + 0$$
$$33 = 3x.$$

"Perfection! Now obliterate!"

He studied the situation. There were 3 numbermasks, so Marco knew he needed to slash everything in thirds.

$$\frac{33}{3} = \frac{3x}{3}$$
$$11 = x.$$

"You've done it! Splendid!"

If Marco's head was a balloon, it would have burst from all the air being pumped inside. He was proud of his accomplishments,

$$2(x - 225) = 498 - x$$

and his hunger for more was beginning to growl in his stomach once again. Mr. Pikake went to the board and began furiously writing. When he finished, he used his long arms to slash everything he had just completed with two broad strokes.

$$18 = 3(x - 5)$$
$$18 = 3(x + (-5))$$
$$18 + 5 = 3(x + (-5) + 5)$$
$$23 = 3(x + 0)$$
$$23 = 3(x)$$

"What was my mistake here, son?" He turned to Marco.

From Marco's hunt, he knew that $3x = 33$, not 23. So, clearly something had gone wrong. He studied the professor's steps, like a general reviewing a previous battle for insights on how to win the war. Then, he saw it. It was a mistake he would never forget. He had made the same mistake before and it was the only time Mr. Pikake had ever become angry with him. "You forced the duel with -5, but it wasn't really -5 because it was under the enlargement spell of the 3. You thought you had vanquished it...but you didn't. Pieces of it still remained."

Marco's explanation led to a loud chuckle from his tutor. "Excellent, son." He tussled the boy's hair, "How could I successfully complete the hunt without the Recta Defensive?"

It was a challenging question. Luckily, all things boiled down to vanquishing or obliterating and Marco knew from the tutor's attempt that vanquishing first in this form would lead to a mess.

"You tried to vanquish first, so I'll obliterate first. Since the 3 is impacting everything on the right, we should just reverse the spell," Marco pronounced proudly. "I'll slash the 3 first, reversing the enlargement...then go in for the vanquish." He divided both sides by 3, or if you will, multiplied both sides by one-third.

$$18 = 3(x - 5)$$
$$\frac{18}{3} = \frac{3(x - 5)}{3}$$
$$6 = (x - 5).$$

$$2x - 333 = 301$$

"Now I have the $x - 5$ alone. So, I only need to vanquish the -5 to truly isolate the numbermask."

$$6 = x + (-5)$$
$$6 + 5 = x + (-5) + 5$$
$$11 = x + 0$$
$$11 = x.$$

The two continued practicing their magic. Mr. Pikake would throw challenging set-up after challenging set-up at Marco, and Marco would slash and obliterate, enlarge and balance, vanquish and hunt again and again.

$$\frac{n}{4} - 3 = 5.$$

"Easy! That one is already unobstructified, just hiding. I can tweak it a bit and use the Recta Defensive."

$$\frac{1}{4}n + (-3) = 5.$$

"Now vanquish then obliterate. I send in 3 to force the duel."

$$\frac{1}{4}n + (-3) + 3 = 5 + 3$$
$$\frac{1}{4}n + 0 = 8$$
$$\frac{1}{4}n = 8.$$

"Now, the numbermask has torn itself into four parts trying to trick me!" With a wave of his marker-wand Marco enlarged the numbermask sending in the reciprocal of 1/4, 4, to do his bidding.

$$4\left(\frac{1}{4}n\right) = 4(8)$$
$$n = 32.$$

"Again Marco, again!" The professor shrieked.

$$x = 53 + \frac{5x}{6}$$

$$7 = \frac{k + 6}{4}.$$

"I have to decide if I want to change the form or not." Marco pondered which way he would prefer. "I see a six, but that is the Numberfolk trying to trick me. The six is being cut into four pieces because the whole thing is cut in four pieces. So, I can't vanquish it yet."

Mr. Pikake leaned in, "Good boy, good. What should we do?"

"I think I'll try this one both ways." Marco first unobstructified to get a good idea of exactly who he was dealing with.

$$7 = \frac{k}{4} + \frac{6}{4}.$$

"Now, I can vanquish the $\frac{6}{4}$. I need to send in its evil twin, $-\frac{6}{4}$.*"

$$7 + \left(-\frac{6}{4}\right) = \frac{k}{4} + \frac{6}{4} + \left(-\frac{6}{4}\right).$$

Marco instantly regretted his choice. The fractions were annoying to say the least. But he was halfway into the battle. A soldier couldn't change his weapon in the middle of combat, neither could Marco. He'd have to continue forward with what he had.

"Ugh," Marco sighed. "I am going to have to convert these all to fractions."

"Indeed." Mr. Pikake stroked his chin, "Indeed."

"Alright, 7 is the same as $\frac{28}{4}$ because 28 divided by 4 is 7."

---

* You might be wondering why Marco didn't simplify the $\frac{6}{4}$ into $\frac{3}{2}$. He could have, but since there are other $\frac{1}{4}$'s in the hunt, he decided to wait 'til the end to do any simplifying of fractions.

$$5(x - 100) = 1095$$

$$\frac{28}{4} + \left(-\frac{6}{4}\right) = \frac{k}{4} + 0.$$

"I went ahead and completed the vanquishing, too. Now, the duel on the left gives me, 22/4. That means..."

$$\frac{22}{4} = \frac{k}{4}.$$

"All that's left is to obliterate. The numbermask has divided itself into four parts, I need to enlarge it to a single mask by multiplying it back to its original size. Dividing by the four is really multiplying by 1/4. So, the reciprocal is 4."

$$4\left(\frac{22}{4}\right) = 4\left(\frac{k}{4}\right)$$
$$22 = k.$$

Marco was surprised, he had done it, the battle was over. He found k and drug the Numberfolk back to base. He was also surprised that k was a Natural. With all the fractions involved, he expected a different result. An itch started attacking Marco's brain. He wasn't sure how to scratch it.

"Very good. What is another battle attack you could take?"

Feeling reluctant and not wanting to deal with fractions Marco groaned as he began the battle simulation once more.

$$7 = \frac{k+6}{4}.$$

The $k + 6$ is really a group formation. The 4 is a suppression spell shrinking the whole thing. Marco needed to enlarge the group back to its original form to be able to hunt k. He started by writing the situation in a more obvious form.

$$7 = \frac{1}{4}(k + 6).$$

"Brilliant!" Mr. Pikake slammed his fist on the table.

$$\frac{4}{5}x = 2^8$$

"I'll send in the reciprocal of the 1/4, a 4 to bring everything back to size."

$$4(7) = 4\left(\frac{1}{4}(k + 6)\right).$$

"I have obliterated the $\frac{1}{4}$, and on the left, that is 28."

$$28 = (k + 6).$$

"We are reduced to a single duel. I'll send in the 6's evil twin to finish things off."

$$28 + (-6) = k + 6 + (-6)$$
$$22 = k + 0$$
$$22 = k.$$

Relief poured over Marco. The impossible itch was gone. "I see it. While unobstructing first made it easier to know what to do, vanquish then obliterate, the battle ended up being harder because I had to deal with fractions."

"Precisely! Every choice has consequences that you must evade. Only you can decide which of the fates is worse." He scribbled a new defensive.

$$48 = 4m + 4(4m - 3).$$

Silence engulfed the room. Marco knew if he sent something in to obliterate, his spell would hit everything. He wanted a smaller spell in his arsenal – did that exist? Deciding he needed to know for sure, he asked, "When I obliterate, can I obliterate just one piece?"

"Expand." This was Mr. Pikake's way of asking for more details.

"Could I send in $\frac{1}{4}$ to only the $4m$?" was the best Marco could do.

$$\frac{(5x + 81)}{9} = 116 + \frac{2x}{9}$$

"*Ahh.* Quite the question. Remember an obliteration charm hits everything. No way around that. You cannot pick and choose. An obliteration will obliterate everything in its path, whether we want it to or not."

The situation was too advanced. There were Numberfolk and masks everywhere. He couldn't tell what was going on. *I have to unobstructify. Then, at least I'll have a clear picture of the battlefield.*

"The $4m - 3$ is together inside the house that has been given an enlargement spell of 4. That means the $4m - 3$ is really $4(4m) - 4(3)$, or $16m - 12$." He wrote his findings.

$$48 = 4m + 16m - 12.$$

"The like terms can be put into a single form. I have four m's and picked up 16 more. So, that is 20 of them."

$$48 = 20m + (-12).$$

"The Recta Defensive! The 20 m's are battling with the -12." Their swords clanked. "I'll vanquish the -12 by forcing the duel with 12."

$$48 + 12 = 20m + (-12) + 12$$
$$60 = 20m + 0$$
$$60 = 20m.$$

"Obliteration is easy now. There are twenty m's huddled together. I need to slash them so only a single m remains. "

$$\frac{60}{20} = \frac{20m}{20}$$
$$3 = m.$$

Mr. Pikake approached the board and started working. Marco saw Merlin, a master wizard, manipulating and wielding his wand with great skill. Explaining his steps as he went, he demonstrated an alternative battle plan.

"The $4m - 3$ are just Numberfolk cloaked together. For my purposes, I will see them as x. I do not know who they are, but

$$\frac{x}{2} - 7 = 154$$

they are someone. Doing this allows me to better assess the situation.

$$48 = 4m + 4(4m - 3)$$
$$48 = 4m + 4(x).$$

"Now, I can see that the enlargement by 4 has been applied to everyone! Well, that simply means..."

$$48 = 4(m + x).$$

"Do you see, son? I may now suppress them down to their true size."

$$\frac{48}{4} = \frac{4}{4}(m + x).$$

"I have obliterated the enlargement on the right!"

$$12 = m + x.$$

"Although it was easy at first to ignore everyone huddled together, I must storm the home now. I must take care of who is inside. I put back the $4m - 3$ where I have x."

$$12 = m + 4m - 3.$$

"Like you, I combine to understand precisely the number of m's that are working together."

$$12 = 5m + (-3).$$

"*Ah*, I too arrive at the Recta Defensive! I complete by vanquishing. I send in the three to duel."

$$12 + 3 = 5m + (-3) + 3$$
$$15 = 5m + 0$$
$$15 = 5m.$$

"A final obliteration is necessary. I must whittle the 5 m's down to only one." He violently slashed at the board.

$$\frac{15}{5} = \frac{5m}{5}.$$

$$6(x - 41) = 4(100 + x)$$

"I obliterated the 5 and thus,"

$$3 = m.$$

Marco stood in awe. How could such a drastically different approach lead to the same conclusion? Then, he remembered one of his favorite things about math – one answer.

"Our paths were different but should always lead to the contemptable charlatan. Whomever the Numberfolk is hiding behind the m… is hiding behind the m. We may uncover him, hunt him, in different ways, but always should find only the one hiding." He turned and smiled kindly at his student. "If you are ever unsure of your results, you may simply place the Numberfolk back into the situation. Test it."

Hoping to one day be as powerful and swift as Mr. Pikake, Marco nodded, trying to take in everything he could. His life depended on it, but more importantly, Marco wanted to keep Maggie safe. He was the big brother, it was his job, and he took it very seriously.

Mr. Pikake sat at the table, the air in the room became thick, Marco could tell there was a shift in his tutor's energy. He motioned for Marco to take the other seat, which he quickly did.

"Marco, I am…" he paused as if trying to find the perfect words, "I am out of time."

He ran his long bony fingers through his hair and stared off into the library. "The SAN requires I name my heir by tomorrow. I – I would love for this to be you."

Marco didn't know what to say. Mr. Pikake was the father he never had. He would do anything for him.

"I'd be honored," he whispered.

"They will test you for your initiation. Do you think you are prepared for such a trial?"

$$-\frac{3}{4}x = 81 - x$$

Standing on a tall mountain, Marco looked over the kingdom below. It was being ravaged by Numberfolk. They needed him to swoop in and save them. Taking a deep breath and stretching a full two inches taller, Marco placed his hands on his hips and looked over the land before charging in. "I'm ready." He said calmly and confidently.

Mr. Pikake pulled out a slip of paper. "A final assignment, to prepare you." He slid the parchment across the table to the boy.

Marco turned the sheet over in his hands. On one side, he saw six battle simulations, on the other an address. "What's this?"

"That is my home. Tomorrow we will meet there. You will be presented with the examination and be inducted into the society. It will be challenging, but you can succeed, Marco. I know you can."

❀ ❀ ❀

Marco returned find an empty house. A message on the refrigerator told him that Maggie was taken to a neighbor to care for her while their mother and Peter went out. He ran to his room to begin his practice right away. He was determined to pass his test. It was the only way to fix things, to make everything he had gone through worth it. Plus, Mr. Pikake was now the only person left in the world who didn't hate him, who he could still make proud.

$$\frac{m+6}{2} = 8.$$

I will obliterate first, Marco thought. He felt daring not using the Recta Defensive and allowing the obstruction to remain, it was exhilarating. *Everything on the left is being cut in half, to return them to full size. I must obliterate using a 2 since 2 is the reciprocal of 1/2.*

$$2\left(\frac{m+6}{2}\right) = 2(8)$$
$$m + 6 = 16.$$

$$\frac{x}{13} - 25 = 0$$

Now, all that is left is the duel. Since the duel is with 6, *I will force a battle with* −6.

$$m + 6 + (-6) = 16 + (-6)$$
$$m + 0 = 10$$
$$m = 10.$$

Remembering Mr. Pikake's advice about how to ensure he had identified the correct Numberfolk, he used the power of substitution. If 10 was hiding, when he tested the scales, they would balance.

$$\frac{10 + 6}{2} = \frac{16}{2} = 8.$$

Marco grinned. He quite liked that trick. The SAN would probably be quite upset if he brought back the wrong Numberfolk from a hunt. This was a nice way to check his accuracy before turning over any information. He moved onto the next simulation.

$$\frac{y}{3} - 10 = -7.$$

Trying to trick me. Marco chuckled to himself. This one, although disguised, was already unobstructed. He rewrote the situation to better see what he was working with.

$$\frac{1}{3}y + (-10) = -7.$$

Vanquish. Obliterate. Since in simple form, I first send in 10 *to duel.*

$$\frac{1}{3}y + (-10) + 10 = -7 + 10$$
$$\frac{1}{3}y + 0 = 3$$
$$\frac{1}{3}y = 3.$$

"Obliterate!" Marco pointed his pencil at the paper as if casting a magical spell. The numbermask had cut itself into three parts in

$$4(x - 250) = 304$$

an attempt to hide. He needed to use obliteration with a 3 to return it to its correct size and conclude the hunt.

$$3\left(\frac{1}{3}y\right) = 3(3)$$
$$y = 9.$$

Wanting to be safe, he decided to check again.

$$\frac{9}{3} - 10 = 3 - 10 = -7.$$

Perfect. Next.

$$4(6 - 5x) = 72.$$

Marco had a choice. He could work out the left to set up for the Recta Defensive or he could go for it. He decided on the latter. There was an enlargement cast on the house of $6 - 5x$, he'd need to reverse that first. He slashed both sides by 4.

$$\frac{4(6 - 5x)}{4} = \frac{72}{4}.$$

Doing the division on scratch paper, 72 divided by 4 was 18. Since he obliterated the four on the left, all that remained was:

$$6 - 5x = 18.$$

This was really $6 + (-5x)$, a duel. *I need to send in -6 to vanquish,* he thought.

$$-6 + 6 + (-5x) = -6 + 18$$
$$0 + (-5x) = 12$$
$$-5x = 12.$$

One last thing to do. The reciprocal of -5 is $\frac{1}{-5}$. I need to obliterate by slashing both sides by -5.

$$\frac{-5x}{-5} = \frac{12}{-5}$$
$$x = -\frac{12}{5}.$$

$$3x + 19 = 1000$$

Marco was concerned. He didn't normally find a fraction, a Logos, hiding behind the mask. He considered hunting x again with the Recta Defensive, but he decided he'd be better off just verifying.

$$4\left(6 - 5\left(\frac{12}{-5}\right)\right) = 4(6 + 12) = 4(18) = 72.$$

It looked good enough, but he still didn't feel satisfied. He wasn't sure he did everything right. He knew subtraction was just adding the evil twin, so subtracting $5\left(\frac{12}{-5}\right)$ really meant adding $5\left(\frac{12}{5}\right)$ and that was adding 12. Everything seemed okay. He decided to go ahead and simulate a different battle attack to be sure.

$$4(6 - 5x) = 72$$
$$24 - 20x = 72$$
$$-24 + 24 + (-20x) = -24 + 72$$
$$0 + (-20x) = 48.$$

The first steps were easy enough. He applied the enlargement to everyone inside, then sent in the evil twin of 24 to vanquish. Now, he somehow ended up with −20 numbermasks. Needing a single mask, he slashed both sides by −20.

$$\frac{-20x}{-20} = \frac{48}{-20}$$
$$x = \frac{48}{-20}.$$

A sinking feeling took over. This was *not* what he found before. What could he have possibly done wrong? In his head he saw the −48/20. Like an FBI facial recognition machine, his brain began running through possible matches. "Oh!" he yelped. The −48/20 was another fraction disguise. He needed to find its true form. He cut both numbers in half to find −24/10 and then again to find −12/5. Marco grinned happily. "Caught you!"

$$4(x - 200) = 2^9$$

He was halfway through his SAN practice. Feeling like he could take on all of Numberville, he continued.

$$68 = -6(-r - 5) + 6r.$$

Marco remembered Mr. Pikake masterfully taking on a similar battle. *No time better than now.* He was safe in his own room, if he was going to try on the grand master hat, now was the perfect time.

"I'll deal with the house of $-r - 5$ later, for now, I will call it H for house."

$$68 = -6H + 6r.$$

Hmm. Both of these appear to be under an enlargement of 6. That means...

$$68 = 6(-H + r).$$

"Now I can obliterate the enlargement by slashing both sides by 6.

$$\frac{68}{6} = \frac{6(-H + r)}{6}.$$

Knowing that eleven sixes was 66, Marco accepted the left wasn't going to be very friendly. He decided to leave it for now and focus on vanquishing what he could before returning to the Logos.

$$\frac{68}{6} = -H + r.$$

"Let's deal with the house." He put the residents back into the simulation.

$$\frac{68}{6} = -(-r - 5) + r.$$

This was a significantly slippery situation. Marco rearranged the players so that:

$$\frac{3x}{7} + \frac{2}{7}x = \frac{329 + 4x}{7}$$

$$\frac{68}{6} = r - (-r - 5).$$

"Subtraction is simply a duel between the evil twin," he mumbled to himself. "I need to find the twins of everyone inside. The twin of $-r$ is r and the twin of -5 is 5, so that means…"

$$\frac{68}{6} = r + (r + 5).$$

The smoke was clearing, the battle was looking possible to conquer now. "I have two r's, so."

$$\frac{68}{6} = 2r + 5.$$

"Vanquishing gives me,"

$$\frac{68}{6} + (-5) = 2r + 5 + (-5)$$
$$\frac{68}{6} + (-5) = 2r + 0$$
$$\frac{68}{6} + (-5) = 2r.$$

Marco wanted to simply slash everything in half and be done with it. But he knew that was harder than it sounded. He needed to take care of the Logos, he grunted.

"Alright, alright. Now -5 can be disguised as $-\frac{30}{6}$."

$$\frac{68}{6} + \frac{-30}{6} = \frac{38}{6}.$$

"That means…"

$$\frac{38}{6} = 2r.$$

Obliterate! Marco was nervous. How could he slash the fraction in half? How did two Vinculums even work? He imagined $\frac{\frac{38}{6}}{2}$ and was disgusted by the look of it. He realized that he didn't need to

$$\frac{(5x - 650)}{5} = 200$$

divide at all. Being so used to slashing, he had forgotten that division was multiplication by the reciprocal, he could instead multiply both sides by $\frac{1}{2}$ to obliterate.

$$\frac{38}{6}\left(\frac{1}{2}\right) = 2r\left(\frac{1}{2}\right).$$

Knowing the 2 would be oblitereated by the $\frac{1}{2}$ and multiplying above and below the Vinculum he found:

$$\frac{38}{12} = r.$$

The facial recognition software started to run, trying to correctly identify the $\frac{38}{12}$. Marco cut both numbers in half:

$$\frac{19}{6} = r.$$

Marco did *not* want to go through the motions to check r's identity, begrudgingly he started in.

$$-6\left(-\frac{19}{6} - 5\right) + 6\left(\frac{19}{6}\right)$$
$$= -6\left(-\frac{19}{6} - \frac{30}{6}\right) + 19$$
$$= -6\left(-\frac{49}{6}\right) + 19$$
$$= 49 + 19$$
$$= 68.$$

Looking back over the battle, Marco was amazed. It was a complex situation, he had done a surprising amount of vanquishing and obliterating, even using Mr. Pikake's method. He felt gross. *How can I ever be a grand master copying another's battle plans? I need to make my own defensives, my own strategies, maybe even ones they will never see coming!* Determined, he tried again. He let the hunt speak to him and allowed his instincts to inform what he would do next.

$$2(x - 222) = 218$$

$$68 = -6(-r - 5) + 6r.$$

"What if I try applying the -6 to the house first? Then I'll get a clearer look at the battlefield and the players involved."

$$68 = 6r + 30 + 6r$$
$$68 = 12r + 30.$$

"I gathered all the r's together and have an unobstructed battlefield. Now, I can vanquish the 30 by forcing a duel with its twin." He felt a little disappointed. Hoping to come up with his own new and unique method had just dropped him right into the Recta Defensive.

$$68 + (-30) = 12r + 30 + (-30)$$
$$38 = 12r + 0$$
$$38 = 12r.$$

"I obliterate, slash both sides by 12…"

$$\frac{38}{12} = \frac{12r}{12}$$
$$\frac{38}{12} = r.$$

"Applying facial recognition," he imitated a robot.

$$\frac{19}{6} = r.$$

Bursting into laughter Marco thought, *That's what you get for trying to be a grand master.* While his attempt using the professor's method took something near thirteen steps, his stab at originality that boiled down to the Recta Defensive was only about six! *SAN can't deny that type of hunting! I was smooth and swift.* His thoughts were interrupted.

"What's so funny?" His mother asked kindly, almost playfully, from the door frame.

Marco jumped. Both her tone and her presence surprised him. "Oh. Er. Nothing. Just doing my math homework."

$$2(x - 211) - x = -2(x - 287)$$

"Laughing about *math*?! That is certainly new. How has your tutoring been going?" She crossed the room and sat on Marco's bed.

"Really good. I sort of see math in a whole new way now."

"That's wonderful, son. I am very proud of you. Your father was marvelous with numbers. I know you will be, too. You are so much like him." She paused and looked down. His mother never talked about Marco's father. He was dying to know more but before he could ask, she continued. "I suppose we should talk about Maggie."

Marco felt like he had been stabbed in the stomach. He didn't want to think about the incident, much less talk about it. On top of that, the fact that his mother was so calm was concerning. "I am really sorry. It was an accident. She was taking my stuff, pulling it from me, and she just, she just fell backwards," Marco lied.

His sister hadn't fallen, he had pushed her, and he knew it. He couldn't bear to admit that he had thrown her. And he certainly couldn't tell her *why* he had done it.

"I assume you didn't break your sister's arm on *purpose*. But honey, this is really serious. I expect more from you. I can't, I just can't understand." She started to cry. Marco felt his heart physically split in two. He thought he was going to throw up.

"I understand. I'm sorry," was all he could muster to keep down his lunch.

"You need to tell her that. But." She wiped her face with her sleeve, "I don't think Maggie is ready to hear it yet. Give her a few days. You need to make amends with her. Family is forever, Marco. She might not show it, but she thinks the world of you, she looks up to you. She...she loves you."

The two sat for a moment in silence, both with their heads hung. Finally, his mother spoke again. "There is something else," her voice buckled, "something else we need to talk to about." *Here it comes*, Marco thought. He was sure this next thing would be about

$$\frac{3x - 123}{3} = 292$$

how he is going to be sent away, his undeniable punishment. "I spoke with your tutor." Marco's head jerked up. What in the world did Mr. Pikake have to do with any of this? "He told me what really happened that day...that day when Peter attacked you." Tears flowed from her eyes like someone had gently twisted open the faucet. "Why didn't you tell me what Peter was doing? Has he been treating you like this? Threatening you?"

Marco was speechless. This is not where he thought this conversation would go. Atop that, he could see his mother's pain. It killed him. Considering everything with Maggie, she didn't need this heartbreak, too. Marco certainly didn't want to be the one to give it to her. He sat like a statue, unsure what to do.

"You have to talk to me, son. You have to tell me things. No one...no one can ever treat you that way. I failed you as your mother. I am so sorry." Her lip quivered and Marco lunged wrapping his arms around her. He wanted to squeeze away every bad feeling. With sheer strength, he wanted to make everything better. "Peter will be leaving. I don't know what will come next, but he can't be here." Marco had the feeling she wanted to say more, but her sadness wouldn't let her. She tucked him into bed and stroked his back until she thought he was asleep and quietly left the room.

To anyone looking in, Marco lay in bed sleeping. But under his blanket, the second his mother shut the door, Marco silently sobbed and sobbed until his eyes were too heavy and too wet to hold open any longer.

HUNTING 101 UNIT 4

$$3x - 942 = 394 - x$$

Here we are. The last blank page. Almost the end of the book but some of the best passages still lie ahead. After much consideration, I have decided to gift this page to you. Fill it up. Choose wisely. Like Marco's journal – make sure everything is just right.

$$8x = \frac{3350 + 30x}{5}$$

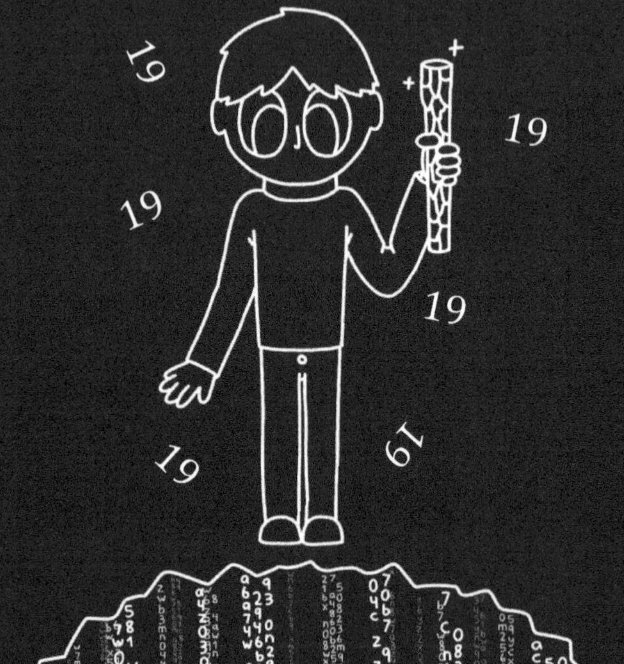
19
19
19
19
19
19

19

ISOLATED

"ACTUALLY, EVERYTHING THAT
CAN BE KNOWN HAS A NUMBER;
FOR IT IS IMPOSSIBLE TO GRASP
ANYTHING WITH THE MIND OR TO
RECOGNIZE IT WITHOUT THIS."

-PHILOLAUS

Searing pain shot through Marco's head. A mixture of the bright sunlight and his dried tears made everything look blurry. Everything was out of control. As much as Marco tried to force himself to believe things were going to get better – Peter would be leaving and wouldn't be his problem anymore – he couldn't make it true. He had so much to fix. He still wasn't right with Oliver, Maggie hated him, Liam despised him, and his mother, well she'd never say it but he knew she'd always blame him for tearing apart their family. He splashed some water on his face in the bathroom before returning to his desk. He had to focus on the only thing he could control – the numbers. Today was his test, his inauguration into the SAN. Having disappointed everyone else in his life, the last thing he wanted to do was to disappoint Mr. Pikake, too.

He stared at the first simulation for what had to have been twenty minutes before he finally dug in.

$$-7 = -2(-4 + 3x) + 2(x + 3).$$

Another complex battlefield, the numbermask had many partners defending it, protecting it. Marco imagined the x, like a king or queen or better yet a president with a swarm of secret agents and decoys surrounding it. Deciding on the known path of the Recta Defensive – unobstructify, vanquish, obliterate – as his headache wouldn't allow for more cunning plans of attack, he started in.

"Okay. The -2 is affecting the entire house of $-4 + 3x$, so let's see where that leaves them." He worked out that a -2 spell on the -4 would result in 8. He enjoyed thinking about the little two-clusters forming in the upside-down and then fusing four of them together in the right-side-up. Similarly, a -2 spell on the $3x$ was really $-6x$. He jotted everything down.

$$-7 = 8 - 6x + 2(x + 3).$$

"Now, for the second decoy. The 2 has been cast on the $x + 3$ group."

$$2((x - 422) - (90 - x)) = 4(106 - x) + 2(x + 290)$$

$$-7 = 8 - 6x + 2x + 6.$$

"Let's round 'em up. I've got -4 x's altogether and then another $8 + 6$ which is," it took him a minute to do the calculations, his brain was moving much more slowly than usual.

$$-7 = -4x + 14.$$

Vanquish. Obliterate. "Sending in the -14 to duel."

$$-7 + (-14) = -4x + 14 + (-14)$$
$$-21 = -4x + 0$$
$$-21 = -4x.$$

"Obliterate the -4." He slashed both sides.

$$\frac{-21}{-4} = x.$$

He didn't have it in him to check. He moved onto the final situation but not before erasing the negatives to write x as $21/4$.

$$-5(n + 1) + 2n = 7(9 - 7n) + 12(1 + 3n).$$

"My gosh!" Marco said aloud. If the last one was a president, this had to be... Marco couldn't even think of a person so important that they would have this kind of security around them.

Not seeing an alternative plan of action, Marco started to weed through the situation. *Okay, there is the 7 enlargement on the house of $9 - 7n$ so that becomes $7(9) - 7(7n)$ which is $63 - 49n$. Then there is another enlargement, this time a 12 on the house of $1 + 3n$ making it $12(1) + 12(3n)$ or $12 + 36n$.*

$$-5(n + 1) + 2n = 63 - 49n + 12 + 36n.$$

Now $63 + 12$ is 75 and -49 n's plus 36 n's is $-13n$'s.

$$-5(n + 1) + 2n = -13n + 75.$$

On to the other side. Applying the -5 to the house of $n + 1$ gives me $-5n + (-5)$.

$$-5n + (-5) + 2n = -13n + 75.$$

$$-3(30 - x) - 2(x + 6) = -3(x - 305) + x$$

Closer, closer. On the left I have $-3n$'s.

$$-3n + (-5) = -13n + 75.$$

I need to get all the masks together. I'll send in $-13n$'s twin.

$$13n + (-3n) + (-5) = 13n + (-13n) + 75$$
$$10n + (-5) = 0 + 75$$
$$10n + (-5) = 75.$$

Okay, now send in the 5 to duel.

$$10n + (-5) + 5 = 75 + 5$$
$$10n + 0 = 80$$
$$10n = 80.$$

Finally! Obliterate. Slash both sides by 10 to get...

$$\frac{10n}{10} = \frac{80}{10}$$
$$n = 8.$$

All that for an eight?! It wasn't at all the intergalactic supreme leader of the Universe he'd anticipated. Marco rolled his eyes before throwing himself back into bed. He felt ready for his initiation, but no way he could impress the SAN in his current state. Pulling his knees to his chest, he squeezed his eyes tightly closed until he fell asleep.

When Marco awoke it wasn't even noon. Still two hours before he needed to head to Mr. Pikake's house. He grabbed his journal and began thumbing through the pages.

Marco was all alone. His family hated him, his friends hated him, too. He hated himself most of all. He fiddled with his journal. Although tutoring had started the slippery slope, this little book sealed his fate. He moaned. Noticing the last page of the journal was peeling back slightly, he started flicking the corner back and forth as he mulled over the last week. He had experienced great highs: his baseball game, Mr. Pikake coming to watch, his progress at becoming a master. He had also experienced some of

$$3(x - 100) + x = 4(x + 100) - 5(x - 200)$$

the lowest lows of his life: fighting with Maggie, hurting her, hurting his mother in the process.

As he continued to pity himself, something caught his attention. The back page of his journal contained a thick card glued to the soft leather binding. He flipped to the front, the card there was thin, a single sheet. Marco examined the back of the book again. It seemed like there were three, no four, sheets on the back while the front had only a single piece. Someone had glued sheet on top of sheet on top of sheet. *Why?*

He carefully pulled back the loose corner to uncover what was hidden beneath. It was blank except for a name written in the bottom left – Bernard Green. He pulled again, the next layer was also blank except for a single name – Miranda Seymour. One more to go, he carefully loosened the sides that hugged each other, bonded by glue, to see the final name – Sarah Snow.

Puzzled, Marco wondered who these people were – why were their names in *his* journal? He grabbed his phone. It was the only thing his mother and Peter hadn't confiscated, yet.

He typed the first name – Bernard Green – into the search engine. News stories filled the screen. Marco clicked on the first and began to read.

Bernard Green, a seventh grader at Buchannan Middle School has been declared missing. His parents went to the local authorities who promptly issued an AMBER Alert on Saturday morning.

Bernard was last seen in the area of his school on Friday. He was wearing a green and white striped sweater and blue jeans. He is approximately 4'10" and 120 pounds with short light brown hair and green eyes. If you have any information or have seen the child, contact your local authorities immediately.

A missing child? Marco stared at the picture attached to the news story of Bernard. He could have been Marco's friend, he was

$$9(x - 9) - 2(x + 20) = 377 + 4(x + 46) + x$$

his age, his smile made him seem so fun and friendly. Just the type of kid Marco would have gotten along with.

He scanned for any updates on the situation and found only one. He clicked the link to reveal a website with a stereotypical green alien and a bigfoot at the top. The page had a black background and yellow text. Marco read the blog's title: *Mysteries and the Unexplained*. He scrolled until he located Bernard's name.

> *Take Bernard Green. All reports point to Bernard being a great kid. In the math club and on the soccer team, Bernard suddenly disappeared in February. There have been no updates to the story. Bernard is the third child in the area to go missing in the last three years. The other two, Miranda Seymore and Sarah Snow are also believed to have been taken by the same suspect. While police have not formally made any statements, our source reports a suspect who they are calling 'The Algebraist' is responsible for all three incidents.*

> *Who is the Algebraist? From what we were able to dig up, the kidnapper with this catchy moniker is part of a secret society – most likely Illuminati – who have been brain farming children. Many suspect the brain farm is a machine that utilizes the special properties only found in the cerebellum of young adults to power a complex algorithm aimed at ultimately using human data through online sources to predict and influence the outcomes of events on a global scale. (Scary stuff.)*

Marco gasped. He quickly switched windows to his messages and started typing to Liam. He stopped. What would he say? Liam had told Marco about the Algebraist long ago – probably when Bernard went missing. He couldn't tell Liam about the SAN and Mr. Pikake, or about the journal with the names, without putting him in danger.

He didn't know what to do. The only person he could ask about the names, the missing children, was his tutor. From what it looked like, Mr. Pikake *was* the Algebraist! Marco began pacing frantically back and forth. Weighing his options.

$$(3x + 1) + 2(x - 8) = 4(x + 85) - 13$$

Should he go to Mr. Pikake's house for the initiation? What if the initiation is when they kidnap the kids? Should he tell someone? If he did, he would be putting them in danger – which was worse. He could pretend to be sick, but what would that do? Delay the inevitable. He was going to have to face this one way or another.

An hour later he had more questions, no answers, and a big flat spot on his rug where he had worn down the threads from his pacing. In a full-on pity party, Marco decided getting kidnapped wouldn't be the worst thing. In fact, it would probably make everyone happier. Peter could return and his mother and Maggie could have the happy little family they all dreamed of – without Marco. There was only one snag. Marco couldn't survive with Maggie always hating him, always thinking of him as the big bully brother she once had. If he was never coming back, he had to at least try to make things right with her. He sat at his desk, dug out a fresh sheet of paper and began writing.

Dear Maggie,

If you are reading this, I am gone. I have been kidnapped. I want to start by saying I love you. I love you so much. You're my sister and can annoy the heck out of me but that is what sisters are supposed to do. You are brilliant and funny and kind, and I love that you are my family.

I know you probably hate me and are glad I've been kidnapped. I didn't mean to hurt you. I've gotten myself into a terrible situation. The journal was a gift from Mr. Pikake – my tutor. He is a part of a secret organization called SAN, the Society for the Abolishment of Numbers. It turns out, numbers are everywhere. They control everything. Once you start to see it, well, it's terrifying. Mr. Pikake has been teaching me to become a master of the numbers. I did it, well, I did it for a lot of reasons. I did it because he was like a dad, he played ball with me, he listened to me, he came to my games and I liked that, I liked that a lot. I also

$$8(x - 7) - 4(x + 13) = (x + 77) + 3(x + 27) - (x - 77)$$

did it for you. When I started to see the numbers, I realized I needed to be a master to be able to protect you, to protect us. Who am I kidding? I'm such a phony. I guess I really did it because I liked how it felt. It made me feel smart and powerful. You are so good with numbers it was nice to finally feel like I could be good too.

The thing is, Mr. Pikake told me I couldn't tell anyone about SAN because it would put them in danger. When you tried to take the journal, I knew you'd figure it out. You're so smart, and I couldn't let that happen. It just turned out that I was the dangerous one. In trying to protect you, I ended up hurting you.

Well, anyway, I am gone now. I went to Mr. Pikake's house to complete my initiation into the SAN. I don't think I want to be initiated anymore. I found out that the journal he gave me belonged to three students before me. When I looked them up they had all gone missing. I am just the latest victim. I probably shouldn't be telling you this, but I couldn't disappear without fixing things with you. Don't go digging Maggie.

Anyhow, you've been a great sister and I have been a crumby brother.

-Marco

He folded the paper in half then again and again until it was so hard and rigid it was impossible to fold it anymore. He taped it together on all sides and wrote

For Maggie's Eyes Only

across the top. He figured eventually she'd find it. She was always snooping around.

He smiled remembering when Maggie was only six or seven, she watched a spy movie motivating her to go around stealing things from everyone. When a neighbor finally complained, Maggie ended up grounded for a month. She tried to explain that

$$2(x - 41) = 3(x + 80) - 2(x - 11)$$

she was a spy and that it wasn't stealing – it was *investigating*. No one seemed to care the difference. Since then, she had only gotten better at 'investigating' without getting caught. She'd always return any items she snatched, quite skilled at both the retrieval and return.

Grabbing his coat, Marco stuffed the journal in his backpack and laid the note perfectly in the center of his desk. He wanted to run to his mother and hug her, but he knew he couldn't.

Making his way downstairs he heard a thud from the kitchen. He pushed his back to the wall, trying to make himself invisible. *What was that?* His mother and Maggie were out, he was supposed to be alone. Clearly, he was not. Another thud made Marco jump. Gripping the rail tightly, he leaned forward as far as he could to try to get a glimpse at what was causing the sound. Then he saw it. The sweat that instantly formed on his palms made him lose his grip of the stairs and he tumbled down to the landing. Peter stood in the kitchen doorway holding a large cardboard box.

Marco stumbled trying to get up. He looked at the front door, it wasn't too far away. If he just made a run for it, he could probably make it.

"Marco." Peter beckoned in a booming voice. He approached his stepson. Marco threw his arms over his face in an attempt to protect himself from the inevitable blow. He waited. It never came. He lowered his arms just enough to peek through and see Peter collecting items and placing them in his box. "You gonna just stay down there?" Peter asked.

"Uh, um, no." Marco stood, readjusting his backpack on his shoulders.

"I guess I should have seen this coming," Peter said in a low tone as he continued to pack. "I'm sorry, Marco. I am sorry for everything."

Marco was sure he was in a dream. This couldn't possibly be happening. He stood frozen. Peter looked over his shoulder

$$3(x + 9) - (x + 99) = 2(x + 101) + (71 - x)$$

locking eyes with the boy. What Marco saw was shocking. The deep brown irises weren't Peter's. They weren't the fiery hatred-filled pits he was used to – they were Maggie's eyes, kind and gentle. "You got a minute?" his stepfather asked.

Nodding Marco sat at the base of stairs while Peter took a seat on the couch across the room. It was sometime before he finally broke the silence. "I have never told you about your father, Marco." He readjusted himself uncomfortably before continuing. "He was a great man. Amazing at everything he did, an amazing friend. Even picked a crappy guy like me to be his best man when he married your mother."

Peter was my dad's friend? Marco thought. Before he could let this news sink in Peter kept going.

"When he died, it broke us all. Your mother and I needed each other just to survive, just to tread water. And when she picked me to continue her life with, I was stunned. I was the luckiest man in the world." He lifted his head to look at Marco but shot his eyes back to the floor immediately, like seeing the boy was physically painful. "You look just like him, you know. A person can't look at you and not see him. I know your mother sees it, too." He buried his face in his hands. "She'd never love me like she loved him. Like she loves you. I was just the consolation prize. And it drove me crazy. Every time she picked you, she was really picking him. It was a reminder of my place, of my importance, of the fact that I'd always be second."

Peter rose to his feet and Marco recoiled, inching back, ready to bolt. His stepfather only turned and continued placing his things in the box. "I thought. Hell, I don't know what I thought. All the time it seemed like the world revolved around you, around your father, and I, I just couldn't deal with it. Now, look where that's got me."

Marco was stunned. Questions roared through his head like river rapids, clashing and banging violently. The beeping from his phone broke the silence. It was his alarm. Only five minutes until

$$10(x + 9) - 4(x + 17) - 63 = 5(10 + (x + 51))$$

his initiation. "You better get going," Peter mumbled. Things were clear. His family would be better off without him. Without the constant ghost of his father lingering over them, they could finally be happy. He knew what he had to do. He wiped a tear from his eye and pushed the front door open, fully expecting never to return.

❧ ❧ ❧

The short stalky man sat in the armchair to the left of the fireplace sipping tea. "Pikake, I have to say the SAN has always been kind to you because of your family involvement, but if this boy isn't the one, well…" he finished his sentence with a loud slurp.

"He's it, he's it, Maxwell. Enough with the threats." Physically, Mr. Pikake towered over the man. But there was something about the power dynamic. It was clear the little man was in charge, and this made the tutor seem somehow smaller.

The knock made Mr. Pikake jump and like a frightened cat, his back curled. He rushed to the next room and flung open the front door. Marco stood on the porch, his head bowed watching the invisible dust he was kicking up with his feet.

"Marco! Son! I am so happy you are here."

The boy looked up and saw his tutor's face. It was warm and kind and *loving*. It wasn't the face Marco expected to see knowing he was a cold-blooded child-abducting brain-farmer.

"Come in, come in. Get yourself warm. I have a fire going." He shuffled the boy into the living room.

Marco began examining his strange environment. The dark figure in front of the fire sent chills down Marco's back. The man was a statue, not even acknowledging Marco's entrance. He looked to his left and studied the huge chalkboard that took up most of the room. On it, he saw an equation. The numbermask m was being drastically protected. Multiple houses were working together. There were other Numberfolk helping, too…lots of

$$5(x + 25) - 3x = x + 2(x + 212) + 3(40 - x) + 275$$

them. Marco examined Mr. Pikake's steps to hunt the m, he slowly and methodically isolated the numbermask. Something about the method seemed...off.

Mr. Pikake ushered Marco to the second chair in the room. As the boy walked around the furniture, the small man tilted his chin just enough to meet Marco's eyes.

"Marco, this is Maxwell Mandel, from the SAN. He will be administering your evaluation."

The voice that came out of the professor was unexpected, it was small and squawky like a mouse. It wasn't the confident over pronounced sounds Marco had become accustomed to. Maxwell gave Marco a slight nod, as if trying to move as little as possible, "Charmed."

Mr. Pikake stood in front of them both, but Marco found his eyes drawn back to the chalkboard. Why would Mr. Pikake solve this in that way? It was brute, forced. It wasn't the work of the grand master he had seen before. There were no bold obliteration steps and cunning rearrangements. It was careful, it was precise. It was like when Marco had tried to be witty and ended up doubling his steps. Why on Earth would his tutor do this?

The room suddenly sucked Marco backwards as his back clung to the chair, he was moving in light speed as he saw the story unfold on the chalkboard. The letters and numbers were changing into people – not Numberfolk but humans. Marco recognized them. It was his mother and Maggie, it was Peter and Liam and Oliver, it was Mrs. Sanders, it was all the people in Marco's life. And there was Marco – m.

This was his plan all along. He hunted me, isolated me, slowly and methodically but he did it. Here I am alone with no one else in the world.

A flashflood of tears rushed his eyes. He pushed them back. He squeezed his eyes shut tight until the dam was strong enough to hold them and then he looked at his tutor. He looked at him with

$$-\frac{1}{2}(m + 148) = 2(50 - m) + m$$

disgust and hatred. He looked at him with bewilderment and shock.

A worried look came over Mr. Pikake's face, "Are you ready, son? Ready to officially become a master?"

"I am not your son!" the words came out much louder and stronger than Marco had intended. They came out with more volume and force than Marco even knew he had within him. All the emotions and stress from the last few days tangled themselves into his words. "Is that what happened to Bernard and Sarah? Is that what happened to Miranda? You made them feel like you cared about them just so you could get them alone?"

In the most movement Marco had seen from Maxwell, he whipped his head around and looked at Mr. Pikake as if ready to pounce.

Not knowing who to address first, Mr. Pikake stepped backwards and raised his hands in defense. His eyes darted back and forth between the two sitting in front of him. Settling on Maxwell, he started "I – I – I didn't tell him anything." Then he turned to Marco, "S- er, Marco. How do you know those names?"

With that, the dam burst open. This was all supposed to be worth it. His friendships were in ruins, his family torn apart. Mr. Pikake, he was supposed to care about Marco, he was supposed to make Marco great. It turned out Marco was just another in a list of broken children. Tears streamed down the boy's face. "They were in *my* journal. The journal you gifted to *me* because *I* was special. They were the kids who came before me, who didn't make it. Who you kidnapped!"

Mr. Pikake dropped to one knee, he grasped Marco's hand. The student recoiled but the professor wouldn't let go. He looked Marco dead in the eyes, "Marco. I would never."

Maxwell stood. He was even shorter standing than he had appeared in the chair. "Boy, Pikake hasn't kidnapped anyone." He held his teacup at his chin and continued sipping, "Yes. The names

$$2(p + 12) + 7(p - 49) - p = 3(10 + p) + 4p$$

you mentioned were being considered for entry into the SAN, but they were unfit. You know, not everyone can see behind the curtain and, well, maintain themselves. They started to see the Numberfolk, they saw them everywhere. They saw them while awake and they saw them when they closed their eyes. They understood the power Numberfolk have, and they couldn't handle it. They broke down."

Marco wiped his eyes, "B-but, I read they went missing! They are gone!"

Turning his back to Marco and Mr. Pikake, Maxwell stared into the fire, "Blame the Numberfolk, not Pikake. They ran, they tried to get away. They quickly realized you can run as much as you'd like, you can even try to hide, the Numberfolk will still be there."

Not believing a word out of the man's mouth, Marco shot back nastily, "You're a liar!"

"I won't argue with you boy, and I won't be called a liar." Maxwell's voice was louder but still calm and controlled. "The Bernard boy entered college this year. Studying something like art, still trying to get away from the Numberfolk," he let out a hearty laugh. "Just wait 'til he finds out how much they are involved with the arts. He will be regretting not letting us train him!"

His eyes had dried up and were starting to burn. Marco looked back at his tutor who still gripped his hand tight. He saw tear drops clinging together in the corner of his eyes. He had the same look he had given Marco when he entered the home, a look of love. "Is he telling the truth?" Marco's words left his mouth with a hard push. They were forceful but also weak and cracked.

"I would never hurt anyone, Marco. I thought you knew me more than that. I thought you..." he looked away. Marco knew the man was sincere. He knew he had let his imagination and his emotions get the best of him. Most of all, he knew Mr. Pikake really cared for him. He felt his muscles tightening as if they were

$$12(y - 12) - 3(2y + 12) = 5(y + 34)$$

working on their own. Not asking his brain for permission like a reflex, he squeezed the professor's hand.

"Are we going to do this or what?" Maxwell finally turned back to the two.

"I think today is not the day, Maxwell." Mr. Pikake said firmly. Standing up to tower over the man.

Marco interjected. "No. No. I'd like to get this over with. I am ready." Speaking calmly, he turned to the professor and forced a smile. His rollercoaster was out of control, there were too many competing thoughts clawing for his attention. *This is how Peter must feel.* Like his stepfather, Marco was flailing, trying to dominate the only variable within reach. For him, that was the numbers. He could control the numbers and he needed to.

With that, the test began. Maxwell reached deep into his suit pocket and presented Marco with a rod. Fashioned from an oak tree, the asymmetrical divots that separated each bark cluster resembled the floor of a desert, dry and cracked, begging for life's liquid. A shiny gleam bounced the light as though the branches' soul had been locked in with a thick lacquer. The rod wasn't pointy but blunt on both ends. Sawed from its mother, the edges were smooth disks with concentric circles like frown lines marking its age at the time of separation.

"What's this for?" Marco asked as he rolled the branch between his hands. Neither man responded. Pulling another, a twin, from his pocket, Maxwell drew a circle with the limb. A cold air burst through the room. Marco couldn't believe his eyes. Inside the circle he saw letters and numbers, millions of them falling from the sky. With another whoosh of his hand, Maxwell had lassoed a handful. Like soldiers, they lined up to create an equation.

$$3(x - 6) + 2(4x + 8) - 23x + 19 = 10(6 - x) - 14(2x + 1) + 28 - 2x.$$

Marco's eyes were wide with intrigue. Fear crept up his chest, a tightness overcame his neck and shoulders.

$$l(2 + 1) = 5(l + 100) - 4\left(l - \frac{101}{2}\right)$$

"Hunt," Maxwell had finished. He sat back comfortably in his chair and lightly sipped on his tea.

"Hunt?" Marco exclaimed. "How?"

"With your rod," Mr. Pikake placed his hand on his student's shoulder. With the other, he grabbed Marco's wrist and began to make tiny circles with the branch. Like a digital board, every move was displayed in the gaping hole Maxwell had created. Marco felt nauseous. Both men stared at him, waiting, it was Marco's move. It was all too much. Wanting to run, to hide, Marco had nowhere to go. He took a deep breath and turned his back to the equation that lay in front of him. Emotionally exhausted, his creativity waned. Not only that, this was the most convoluted situation he had ever seen. Another sharp inhalation. He'd use the Recta Defensive, it was a sure-fire method and a safe bet.

Holding his arm out straight, the rod an extension of his hand, an extension of himself, Marco shouted, "Cast the enlargement spell on the house of $x - 6$!" His hand waved back and forth to reveal:

$$3x - 18 + 2(4x + 8) - 23x + 19 = 10(6 - x) - 14(2x + 1) + 28 - 2x.$$

"Again! Enlarge the house of $4x + 8$!" Another few flicks and the $2(4x + 8)$ disintegrated, unveiling $8x + 16$ in its place.

$$3x - 18 + 8x + 16 - 23x + 19 = 10(6 - x) - 14(2x + 1) + 28 - 2x.$$

Marco swirled his hand circling the $3x$, $8x$, and $-23x$. He threw his hand to the left, across his body and the terms obeyed.

$$3x + 8x - 23x - 18 + 16 + 19 = 10(6 - x) - 14(2x + 1) + 28 - 2x.$$

"Duel!" he shrieked. He was overcome with excitement. He was a conductor composing his own tune, commanding the Numberfolk to do as he demanded. And they followed his every word. They awaited his order. In his head, Marco imagined the fireworks. Thinking of Maggie, a smile spread across his face. The 19 fractured into positive singletons while the -18 fractured into negatives. They exploded in the sky leaving only one positive

$$5x - 4(x + 36) + (x + 5) = -4(x - 52) + 5(x + 1)$$

charge remaining. Pushing that together with the 16, he slashed a one seven into the hole.

$$3x + 8x - 23x + 17 = 10(6 - x) - 14(2x + 1) + 28 - 2x.$$

Now it was time for the like terms, numbermasks all under proliferation charms. He pushed the $3x$ and the $8x$ together revealing eleven x's. He sent them up against the $-23x$. Explosions fired off leaving -12 masks as the dust settled.

$$-12x + 17 = 10(6 - x) - 14(2x + 1) + 28 - 2x.$$

He turned his focus to the right. "Enlarge the house of $6 - x$!" Six slashes and circles resulted in $60 - 10x$ in its place. "Again, the -14 on the house of $2x + 1$!" More dips and dives with his wrist. He looked at the result, proud of how far he had come.

$$-12x + 17 = 60 - 10x - 28x - 14 + 28 - 2x.$$

Circling the $-2x$, he flicked it to the left followed by a quick flick of 60 to the right.

$$-12x + 17 = -10x - 28x - 2x + 60 - 14 + 28.$$

He chuckled out loud at how simple his next step would be. "-40 numbermasks!" Marco stabbed the rod forward as the x terms broke apart and rebuilt themselves as he directed.

$$-12x + 17 = -40x + 60 - 14 + 28.$$

"Duel!" His voice was becoming louder, more confident. The 28 and -14 erupted together fourteen positive particles floating down from the wreckage. He quickly summed the 60 and the 14, standing up straighter as he saw he had broken down a complex camouflage system into four measly terms. Whatever Numberfolk was hiding didn't stand a chance. Marco was ready to rip apart its disguise.

$$-12x + 17 = -40x + 74.$$

A mad scientist with his potions bubbling before him, a wicked smile simmered across his face. Violently stabbing at the hole, he

$$10(2k - 6) + 19(4 - k) - 2 = -k + 3(k + 123) - 2(k + 1)$$

screamed, "Send in −17, the evil twin." He didn't even need to motion, the −17 appeared before him.

$$-12x + 17 + (-17) = -40x + 74 + (-17).$$

He twisted and turned, flipped and bent. "Vanquish!" With two quick jabs, the 17 was no more. It melted away in front of him.

$$-12x + 0 = -40x + 74 + (-17)$$
$$-12x = -40x + 74 + (-17).$$

A horrible feeling filled his gut. The subtraction on the right was more challenging than Marco wanted it to be. Scared the men would notice his hesitation, he decided to count. He felt powerful when he counted and could use that burst of energy at this exact moment. He counted three to get to twenty, then fifty to arrive at seventy. Four more got him to seventy-four. All together that was, "57!" he yelled as he slashed.

$$-12x = -40x + 57.$$

This was it. He was almost there. He threw −40x's evil twin into the hole.

$$-12x + 40x = -40x + 57 + 40x.$$

"Vanquish!" he was laughing now. The power had completely taken over. He jumped over two steps to uncover:

$$28x = 57.$$

"Obliterate!" The words bounced around the tiny living room. Out of the corner of his eye, Marco saw Mr. Pikake on the edge of his seat, leaning forward, the bright sparks from the circle's edge reflecting in his blue eyes.

Marco slashed both sides by 28.

$$\frac{28x}{28} = \frac{57}{28}.$$

He hesitated. He knew he could leave no stone unturned. Hunting x was good, but it meant nothing if x still wore a mask.

$$2\big(3(x + 6) - 2(x - 16)\big) = 2(x + 195) + (64 - x)$$

Quickly running through the possibilities, he tore apart each Integer in his mind. The 57 became $3 \cdot 19$ while the 28's soul was nothing more than two 2's and a 7. Knowing his work was complete, he flicked his wrist to finish his hunt.

$$x = \frac{57}{28}.$$

A jail appeared, a cage. It came from what seemed like an abyss. A crocodile stalking its meal, its jaw widened as it inched towards the face of the hole. The $\frac{57}{28}$ was running. It became larger and larger as it approached the face of the great hole. Maxwell sprung to his feet. Waving his hand in the same circular motion but clockwise this time, the hole began to shrink. Marco wasn't sure it was enough. The $\frac{57}{28}$ was still coming. The five and the seven held each other, giggling manically. Stepping back, Marco tripped on the rug. His fall didn't stop him. He continued to push himself backwards away from the hole, away from the five. At the last moment, the cage caught the Logos. Its jaws clamped shut locking it in and the hole disappeared as quickly as it had arrived.

Marco collapsed his full body to the floor and sighed in relief. Maxwell stood over the boy. "Nicely done. You have passed your examination. Welcome to the SAN, Marco." Snatching the rod from Marco's hand, he turned away and began collecting his things. He went on to announce that Marco would be given the rank of Aspiring Master, and in the coming days he would be taken to the SAN headquarters and be shown around. He placed a patch on the table to commemorate Marco's accomplishment before turning to Mr. Pikake and directing him to continue the boy's training. With that, Maxwell crossed the room and disappeared. Not saying goodbye to anyone, it was like he had never been there at all.

Marco and Mr. Pikake sat in silence until finally the tutor spoke, "I suppose congratulations are in order?" He stood and walked out of the room. Marco half expected the professor to disappear

$$x - 4(x + 4) = 5(71 - x) - (16 - x)$$

like Maxwell and was surprised when he returned, holding a cake. "I knew you would do it," Mr. Pikake whispered kindly.

The cake was chocolate with bright yellow frosting. Across its face the words 'Congratulations Son!' sprawled in black cursive. Marco burst into tears. He felt like he had cried more in the last twenty-four hours than he had in his entire life.

"I am so sorry I doubted you. I just, I found the names in the journal and when I looked them up, I didn't know what to think. Plus, all my friends are mad at me, my family hates me. I…I'm just falling apart."

Mr. Pikake placed his hand on Marco's shoulder, "It's okay. I am sorry too, Marco. I am sorry I made you think I could ever do such a thing. I care for you, Marco. You are the son I never had. The son I *wish* I had. Things will work themselves out. I am sure of it."

"I know that equation," Marco motioned to the chalkboard. "It's about me, isn't it? You were hunting me, trying to get me alone. Why couldn't I tell my sister about this? Why all the secrets?"

"*Ahh*. Yes. I am impressed you figured that out. The equation – it is you Marco, but I was not *hunting* you. I have been in the SAN nearly my whole life. Math, equations, it is all I know. I am not great with people. The equation is my way of determining how to bond with you, my way of uncovering who you are. We all wear masks, we only let people see what we want them to. I needed to get behind your mask, to understand you. That is what this is."

Feeling like he had been hit with a bag of bricks Marco sat quietly trying to regain his breath. His rollercoaster had worked overtime today. He went through isolation, shock, fear, and acceptance. Through anger, relief, and celebration. Like a kitchen sink soup, Marco was swimming in an emotional hodgepodge. But now, now Marco tasted a different ingredient. One he couldn't quite put his finger on. As the two ate cake in silence, Marco's brain was firing on all cylinders to figure out what this new feeling was.

$$16(a - 2) + 5(6 - a) = 4(a + 81) - 6(-5 - a)$$

Marco entered his house to find it dark. Peter must have finished packing and his mother and Maggie were still out. He felt relieved. He didn't want to talk to anyone, he just wanted to collapse onto his bed and let sleep wash away his emotions. With the way he felt, he could sleep for a week. He reached the top of the stairs and grabbing the door frame for leverage he swung into his room. Maggie sat on the edge of his bed. When he caught her eyes, he was surprised to see compassion. All he had seen from her recently was anger and what he could only describe as shame. As he looked at his sister, he noticed she had something in her hands. In an instant he recognized the writing on the back.

"We need to talk," she said softly.

Marco's Hunt

$$3(x - 6) + 2(4x + 8) - 23x + 19 = 10(6 - x) - 14(2x + 1) + 28 - 2x$$
$$3x - 18 + 2(4x + 8) - 23x + 19 = 10(6 - x) - 14(2x + 1) + 28 - 2x$$
$$3x - 18 + 8x + 16 - 23x + 19 = 10(6 - x) - 14(2x + 1) + 28 - 2x$$
$$3x + 8x - 23x - 18 + 16 + 19 = 10(6 - x) - 14(2x + 1) + 28 - 2x$$
$$3x + 8x - 23x + 17 = 10(6 - x) - 14(2x + 1) + 28 - 2x$$
$$-12x + 17 = 10(6 - x) - 14(2x + 1) + 28 - 2x$$
$$-12x + 17 = 60 - 10x - 14(2x + 1) + 28 - 2x$$
$$-12x + 17 = 60 - 10x - 28x - 14 + 28 - 2x$$
$$-12x + 17 = -10x - 28x - 2x + 60 - 14 + 28$$
$$-12x + 17 = -40x + 60 - 14 + 28$$
$$-12x + 17 = -40x + 74$$
$$-12x + 17 + (-17) = -40x + 74 + (-17)$$
$$-12x + 0 = -40x + 57$$
$$-12x = -40x + 57$$
$$-12x + 40x = -40x + 57 + 40x$$
$$28x = 57$$
$$\frac{28x}{28} = \frac{57}{28}$$
$$x = \frac{57}{28}.$$

$$3(y - 9) - 4(5 - 2y) - 2y + 16 = 2(5y + 46) - 3(-y - 74) - 5y + 12$$

$$10^2 \left(\frac{1}{10} + \frac{1}{10} \right)$$

$$5 \cdot 2^3 / 2$$

$$2^4 + 2^2$$

$$\sqrt{400}$$

$$4 + 5 + 5 + 6$$

$$2 \cdot 5 \left(\frac{1}{5} \right) 5 \cdot 2$$

20

GRADUATION

"THAT AWKWARD MOMENT WHEN YOU FINISH A MATH PROBLEM AND YOUR ANSWER ISN'T EVEN ONE OF THE CHOICES."

-RITU GHATOUREY

There was no escaping it, Maggie had read Marco's letter. He hadn't expected her to find the note so quickly but leave it to his sister to be ahead of the timeline. Their talk began with Marco spilling his guts. He told her everything. He told her about Mr. Pikake, about the SAN, about his initiation and becoming an Aspiring Master, about the missing children, about his fights with his friends, and most importantly, he told her about the numbers. Their talk ended with Peter. While Marco expected Maggie to blame him for the loss of her father, he was surprised that she blamed herself. She felt horrible that Marco had been treated so badly, and it wasn't like she'd never see him again, he already had plans to show Maggie his new apartment next weekend.

Maggie wasn't surprised at all about the fact that Numberfolk controlled all of reality. She went on a thirty-minute tangent about a movie she saw, *The Great Math Mystery*, that had explained how mathematics was the language of the Universe. "They even said that the world is math – that is like all it is. Like *all*. Everything is numbers, like *The Matrix*."

Marco had never seen *The Matrix*, so he couldn't really follow along. He was astonished by his sister's ability to grapple with this information and run with it. Together, they created a plan, it would take a lot of rebuilding on Marco's part, but it could be done. Liam would be their first target – he would be the easiest to sway. They were right. After Marco apologized dramatically and verified his crazy conspiracy theories it was like a winning lottery ticket for his friend. Hearing that numbers control everything wasn't at all terrifying for him – it was freeing. He always knew there was some unseen force behind it all. The fact that there was a secret society was even better – the icing on the cake.

Oliver was more challenging. He was smart and logical. Marco played to his vanity explaining how they couldn't do it without Oliver's math know-how. Marco's admitting of Oliver's mathematical superiority was just what was needed. Although Marco suspected it was more FOMO than anything that eventually got him on board. It was done. Together, Marco,

Maggie, Liam, and Oliver had formed their own organization, now all they needed was a name.

"I think we should call it Wolfram," Maggie chirped as the three huddled in Marco's room on Monday night. "Wolfram is the strongest element – it will strike fear in the hearts of anyone who hears it! They'll know we mean business. Plus, Wolfram is like a super powerful math AI, so it is fitting."

"Are you crazy?!" Liam looked like he was about to burst. "You think Wolfram isn't already in cahoots with the SAN? They probably have a vast network and are embedded in anything you can think of. I thought you were supposed to be smart. Assume they are everywhere. That is Secret Society 101."

"What about KAN – Kids Against Numbers?" Oliver suggested, "It's cool because we can say things like 'We KAN' and it has like, a double meaning." A mixture of laughter and moaning rang out from the group.

"Are we really against numbers?" Marco spoke up. "I mean, I know that they are everywhere, controlling everything, but is going up *against* them the right thing? The way Mr. Pikake told me about them, they, they were like people. They had communities, games, they worked together to help each other. Don't we want to better understand them before launching a full-scale war on the Numberfolk?" The others chuckled. They might have been on board to form their own society and particularly to help their friend, but they weren't quite ready to pick up the SAN terminology.

"Marco's right," Maggie chirped. The shock nearly made her brother fall over. He felt sure those words had never before left her lips. "Our goal should be to become masters, like Marco. To understand the er-Numberfolk, and how they control the world… why they do it. Most wars are started because of an intolerance to comprehend another's perspective. We can change that. We *should* change that."

"KFUN" Marco muttered. It came out as one word – kuhuffin.

"What's that?" Oliver chimed in.

"KFUN, Kids For the Understanding of Numbers."

"Not as witty as my suggestion, but I like it!" Oliver exclaimed. "Can we pronounce it as Kay-Fun, rather than kuhuffin? We might have a better chance at recruiting others that way." They all burst into laughter and with that, KFUN was born.

They concluded their first official meeting. While Marco felt hopeful, something was still bothering him. It took him until Thursday, his next baseball game, to realize what it was.

When he stepped onto the field and saw his mother and Maggie in the front row – like always – a wave of relief came over him. He couldn't believe everything that happened in just a week. Last Thursday, Maggie didn't have a cast, she didn't know about the SAN, Peter was an ever-looming threat, and a few days ago Marco would have thought they'd hate him forever and never be at a game again. Yet there they were, yelling and cheering for him like nothing had changed. As he scanned the bleachers another face surprised him, Mr. Pikake. Alone in the top left corner of the stands he sat rooting for Marco.

That night, he snuck into Maggie's room after their mother was asleep.

"We are just a bunch of kids. And while I am better – much better – at math, we haven't even started Algebra yet. Not to mention Geometry and Calculus when we get to high school. How are we supposed to all master the numbers on our own?"

Maggie sat silent for a moment, thinking. "Right again, brother." If nothing else, Marco was starting to adore this new Maggie. A Maggie who seemed much more willing to give him credit when it was due. "We need someone on the inside. It's not just the numbers we have to worry about, it's the SAN, too. We have no idea what they are capable of."

Perfect, Marco thought. Knowing he'd have better luck if she thought it was her idea he tossed a soft ball, "Yeah. Who could we get on the inside? I mean, we don't even know anything really about how they work."

"What about your tutor?" she asked. It took everything in Marco not to let his smile take over his face. Just as he wanted, she had taken the bait.

"That's a great idea!" He tried to sound genuine. "Mr. Pikake really cares about me. He'd want to help, and he knows everything about the SAN." Maggie agreed much more willingly and quickly than Marco had expected. Together they hatched a plan to bring the tutor into the fold.

* * *

Marco tapped his foot impatiently. In their study room at the library, he glared at the clock. Like waiting for water to boil, as the seconds ticked away, time seemed to be moving much more slowly than normal. He flashed his tutor an enormous smile when he finally entered the room.

"How is our new Aspiring Master doing?!" Mr. Pikake gleefully sung.

"Good, good. Very good." Marco responded.

"Are you ready to begin the next phase of our work? Now that you are adept in number hunting, it is time to broaden our scope to investigate the real battlefield…the coordinate plane."

"I-I was actually hoping we could get to know each other a little better? You know, just talk?" Marco softly suggested. "With all that happened last week, well – er – I think it would be nice to start over a bit."

"Ah." The tutor sat his briefcase on the table and slid into the chair. "Tis an excellent idea. But." The familiar pop was endearing to Marco. "I have never had someone to share my tales with. I don't know that I would be very good at it."

"I'm sure you'll be great!" Marco's words came out phony and with more excitement than he had intended. "Maybe we can start with your family? You know all about mine." His determination got the best of him as a flood of questions came hurling out. "Can you tell me about your family? How did they get involved with the SAN? What exactly does the SAN do? Do you have brothers and sisters that work for them, too?"

"Slow down, slow down," Mr. Pikake gave a friendly chuckle. "I suppose it is quite natural coming from a Slant family."

There was that word again, *Slants*. Marco thought of it as a bad word, a word that looked down on non-math people, people like Marco used to be. He couldn't get distracted, he needed to complete his own defensive. "Will you tell me about it?" Marco leaned in.

After fiddling with his hands for a moment the professor began. "I believe I already shared with you my early love of numbers. I was about seven when my father began indoctrinating me with the teachings of the SAN. You see, well, the main branch of the SAN is to discover the patterns in the world and identify not only the Numberfolk controlling them, but to understand how they work and to thwart them. To create inventions to keep humans safe and maintain as much control as we can. They ultimately want a world where humans dominate, not Numberfolk. To see behind the curtain and overtake all that is there."

"But if you love numbers, why would you work for a society that wants to *abolish* them?"

Mr. Pikake pushed all the air out of his lungs in a strong sigh. "I – I had an older brother, Fredrick. When our father introduced him to the SAN, he rebelled. Like me, Fredrick loved numbers. Not only did he love them, he was fantastic with them. He refused to be a part of the society, he didn't want to abolish Numberfolk, he wanted to understand them, work with them. The whole idea of it disgusted my father. He couldn't understand willingly submitting to being controlled. Our father disowned Fredrick... if he wasn't a part of the society, he wasn't a part of our family. Then, one night, Fredrick left. I remember it was right before

Christmas. I haven't seen him since. I – I don't know if he is even alive." He bowed his head and after a short pause, continued. "I was our father's only remaining heir. You have seen how important this is to the SAN. As more and more people develop a hatred of mathematics, the SAN loses power. They require now that every Saint – member of the SAN – produces an heir, a replacement for when they are gone. I didn't have a choice. I am the only one left to carry the torch, so to say."

This was it, this was Marco's in. He knew first-hand the power of a sibling, the love that existed whether you wanted it to or not. "I think I'd like Fredrick very much." Marco spoke carefully.

Mr. Pikake's Cheshire grin engulfed his face, "Oh you would. He was simply wonderful. And I am sure he would be a much better teacher than I am."

Looking straight into his tutor's eyes that gleamed with kindness, Marco saw just what he needed...a tinge of despair, of regret. "I don't want to be in the SAN" he started. "I want you to teach me, to teach us, all about the Numberfolk, all about the patterns. We want to learn, but not to abolish them."

"We?" A surprised look came over Mr. Pikake's face. Marco took something out of his backpack and pushed it across the table. The tutor picked it up and studied it. It was a picture. A photo of Marco, Maggie, Liam, and Oliver holding the sign they had created that read 'Kids for the Understanding of Numbers'. A warm expression melted over Mr. Pikake's face. "It's beautiful," he said before setting the photo down and grasping Marco's hand. "You don't know what you are up against, son. The SAN, they are powerful – connected. They have been working for hundreds of years collecting the knowledge of Numberfolk they need. They are driven by fear, and fear can make people do despicable things."

"That is why we need you! We can't possibly do it alone. We need a guide. We need someone who understands Numberfolk and the SAN. It – it has to be you."

$$71 + 72 + 73 + 74 + 75$$

Marco could see the gears turning in Mr. Pikake's head as he considered the proposition. Knowing this would either be a miracle or a disaster, Marco held his breath. If his tutor agreed, they could become a great force – they had a chance. But if he didn't, they could be broken down before they even begun. He could report them to the SAN, it would be horrible.

A wicked grin slowly walked up the left side of Mr. Pikake's face. "We'd have to be careful. We'd have to be very secretive. Most of all we'd have to pretend to still be on their side, work in the shadows."

With a smile, a nod, and a handshake, Marco welcomed Mr. Pikake into KFUN.

❖ ❖ ❖

"So that's it?" Maggie was bouncing on Marco's bed.

"That's it," he replied with satisfaction. "He is going to help us. But we can't let the SAN catch on. We'll have to be careful."

"I can be careful!" Maggie bobbed, holding her palms together pretending to be a secret agent.

He smiled at his sister's antics thinking back to Mrs. Lucas' fortune. He realized the patterns had tricked him. It wasn't talking about Mr. Pikake at all, it was about the SAN. The society was very old, but its entrance into Marco's life had unlocked the trustiest thing of all – family. Not just Maggie but Liam and Oliver too, and Mr. Pikake. A warmness rushed over Marco. It was the feeling he couldn't quite place as he ate cake with his professor. Although still not sure what it was, he allowed it to come, and he bathed in it. Whatever it was, it felt good.

As the two laughed and plotted, neither of them noticed the shadow. Just out the window, behind the shrubs that lined the street a man stood watching their every move.

"What's next?" she vaulted off the bed and landed next to Marco's desk.

121 + 122 + 123

"We will get you caught up. You have to learn all about the history of Numberville, the numbermasks, and hunting. Mr. Pikake said that next we will study the battlefield. Er- coordinates is what he called them." As Marco spoke, they started to appear. He saw them everywhere. His room was covered in horizontal and vertical lines. Everything had two numbermasks attached to it. To top things off, little equations bobbled up and down off every surface in the room. *There's not enough tape in the world*, Marco thought.

"You know what we are?" Maggie bent over, her elbows on the desk and her face so close to Marco their noses almost collided snapping his focus away from the invading Numberfolk.

"What are we?" he dipped his head forward to tap her nose.

"Math kids. We are Math kids." With a giggle she thrust herself forward to headbutt her brother who quickly retaliated tackling her onto his bed.

Together they laughed and they laughed. They laughed so much it hurt.

In the whole history of the world, there has never been an over-the-top teenager plot that worked out as expected. Despite the best intentions, planning, and all the hope and optimism childhood brings, the pollyannaish and ingenuous nature of youth makes them unsuspecting and unable to anticipate what was sure to come next. For Marco and his sister would find themselves part of a tumultuous tale of danger and intrigue and of course, numbers… lots and lots of them.

AFTERWARD

So now you know the tale. Or at least the beginning – how it all started. Marco, Maggie, Liam, Oliver, and Mr. Pikake have their own adventure that has already begun to unfold. But (we don't pop), before you go, we wanted to leave you with a few key points.

Almost everything in this story is completely true. Yes. Numberfolk exist. They really are magical creatures that control the world, that control everything you see and everything you hear. They control everything around you and everything inside you. Also $0.9\overline{9}$ does equal 1.

But more importantly, it is very true that every day some student is labeled as 'not a math kid'. It is not at all true however that there is such a thing as a 'math kid'. Which makes it mathematically impossible to be 'not one'.

The understanding of the Numberfolk and their power is waning, and we don't even want to think about what might happen if we drop that ball. So before you leave, we urge you to become your own master. The next page contains the mathy terms for what Marco learned. Learn them too – become a wizard because there will always always be mathematics. ∎

TOPICS COVERED IN THIS BOOK

- ORDER OF OPERATIONS

- FACTORS AND MULTIPLES

- PRIME AND COMPOSITE

- INTEGER OPERATIONS

- RATIONAL OPERATIONS

- VARIABLES AND EXPRESSIONS

- LIKE TERMS

- SUBSTITUTION

- SIMPLIFYING ALGEBRAIC EXPRESSIONS

- SOLVING ONE-STEP EQUATIONS WITH ADDITION AND MULTIPLICATION

- SOLVING MULTI-STEP EQUATIONS

- SOLVING INEQUALITIES

Be on the lookout as the journey continues in Marco the Great and the Mystery of Phaseville!

Acknowledgements

$\mathbb{Z}$: To the Integers – those who came first, paved the way, and are the basis for everything that would come. Don, Abi, and Elizabeth, my grandparents, all passed. I know how proud you would be, and I thought of you constantly as I created this world. Thank you for helping mold me into who I am today.

$\mathbb{Q}$: To the Rationals – the common denominators who care only that we make good choices, live up to our potential, and remind us that we are never alone. My parents; Charles, and Debra. The power of having someone in your corner, someone who is always there to pick you up when you are down, who always believes in you, and always insists you can do anything you put your mind to is Earth shattering. Thank you for every ounce of love and support, for all your patience, and trying desperately to keep me rational. I was also lucky enough to pick up a third parent, my stepfather Joe – who has always treated me like his own.

$\mathbb{R}$: To the Reals – the friends and colleagues and people who have stumbled into my life and made an everlasting change. To Carla, from the second we met she lit within me a fire of inspiration that turned my mess of ideas into an actual story. To Erich, who always makes math interesting, is a true friend, and generously devoted his time to Marco's world. To the friends and colleagues who have supported, inspired, and cheered me on – there are far too many of you to name, but you definitely know who you are, and I am forever grateful we found our way into each other's lives. And to Mr. Holmes, who made me believe I could write and made me seriously consider every single comma.

ℂ: To the Complex – the perfect dose of imagination that makes each day an adventure. To my *i*'s, the ones *i* wake up to each morning and who hug me each day. To my husband, together we discovered infinity and have never looked back. Without your daily support, tireless reading, listening as *i* share ideas, and belief in me, this would never have become a real story. *i* love you more than words can do justice. To our wonderful children. To our baby who has spunk and attitude and more personality than her tiny body can contain – *i* love you dearly. To our sons – the ones who say, "mom can do anything" and truly believe it. Your love of numbers inspires me, and *i* absolutely love watching you blossom every day. And even though it's not mathematically correct – *i* love you ∞ + 1. To our eldest daughters, you have grown into beautiful people. You are unique, funny, kind, loving, smart, and determined. *i* could not be more proud to be your mom. *i* cannot thank you enough for all the time and love you poured into this tale. *i* see you in Marco and in Maggie, in Oliver, and in Liam. All their best parts were inspired by you. *i* love you the biggest much.

Without each of you, my set would be incomplete.

Thank you.

And to Spiderman who said, "You know what's cooler than magic? Math!"

SK Bennett is an award-winning educator, instructional designer, mathematician, and mom of five. When she witnessed how much joy and understanding stemmed from the relatable analogies and storytelling techniques she used in the classroom, she was determined to bring these same opportunities to a wider audience. Bennett truly believes that all children are 'math kids' and hopes every student can see a bit of themselves in Marco to discover and perfect their own magical powers.